ip in Education
Engagement

T... ology
o... Isles

General editors:
Eric H. Brown and Keith Clayton

Ireland

In the same series

Scotland   *J B Sissons*

Northern England   *Cuchlaine A M King*

In preparation

Wales and Southwest England
Eric H Brown, D Q Bowen & R S Waters

Southeast and Southern England   *D K C Jones*

Midlands and Eastern England
Keith Clayton & Allan Straw

Ireland

G. L. Herries Davies & Nicholas Stephens

*with contributions on
the Pleistocene history
from* Francis M. Synge

Methuen & Co Ltd

First published in 1978
by Methuen & Co Ltd
11 New Fetter Lane, London EC4P 4EE
© 1978 G. L. Herries Davies and Nicholas Stephens

Typeset by Red Lion Setters, Holborn, London
Printed in Great Britain
at the University Press
Cambridge

ISBN 0 416 84640 8 (hardbound)
ISBN 0 416 84650 5 (paperback)

In memory of
Anthony Farrington 1893–1973
Teacher and Friend

Contents

General editors' preface   ix
Preface   xi

1 GEOLOGICAL BACKGROUND   1

2 ROCKS AND RELIEF   11

3 GEOMORPHIC REGIONS   17
The Central Lowland 17
The Leinster Axis 24
The Ridge and Valley Province 32
The southern mountain inliers 39
The Castlecomer and Slieveardagh Plateaux 40
The Abbeyfeale Plateau 41
The Listowel — Killorglin Lowland 42
The Clare Plateau 44
The Iar-Connacht Lowland 46
The Killary Mountains 48
Northwestern Mayo 51
The Ox Mountains inlier 54
The Donegal — Ballina Lowland 55
The Cuilcagh Plateau country 56
The Donegal Highlands 59
The mid-Ulster Highlands 64
The basaltic plateau 66
The Cavan — Down Hill country 72
The Tertiary igneous mountains 73

4 TERTIARY HISTORY 79
The island's primordial surface 80
Evidence of large-scale Tertiary denudation 83
Possible Tertiary diastrophism 84
The planation surfaces 90
Low Plio-Pleistocene sea-levels 98

5 DRAINAGE PATTERNS 100
The rivers of Munster 103
Rivers in the uplands of Leinster 107
Rivers in the uplands of Connacht 110
The rivers of Ulster 111
Rivers in the Central Lowland 113

6 PLEISTOCENE EVENTS 115
The Munsterian glaciations 116
The Midlandian glaciations 118
Interglacial deposits 123
The 'pre-glacial' beach and associated Head deposits 124
Pleistocene geomorphology in the four provinces 125

7 COASTLINE 181
Leinster: Counties Louth, Meath, Dublin,
Wicklow and Wexford 183
Munster: Counties Waterford, Cork, Kerry,
Limerick and Clare 189
Connacht: Counties Galway, Mayo, Sligo
and Leitrim 195
Ulster: Counties Donegal, Londonderry,
Antrim and Down 199

References 222
Index 243

General editors preface

The British Isles is a most varied and complex part of the world, and over a long period it has been studied in great detail by geologists and geomorphologists. Several general studies of the geology and scenery type have appeared, but despite the wealth of the published literature, more detailed work has been limited to a few regions. Indeed although the most ambitious work, Wooldridge and Linton on Southeast England was first published in 1938, no comparable volume has covered any other part of England.

The main difficulty facing anyone writing about the land forms of the British Isles and their history is the paradox that despite a very considerable literature, the gaps and uncertainties seem to diminish very little over the years. It is often said that each piece of research exposes half a dozen new problems, and so far most of the published literature has tended to add to the complexity of our knowledge. The only way out of this problem is through the establishment of more general concepts into which this detail will fit, and this is the role of this series of books on the different parts of the British Isles. The synthesis is certainly not easy, and it is obviously incomplete. Our understanding of the present land surface and its relationship with the underlying geology (whether 'solid' or 'drift') is relatively secure, but the historical background to the present, complicated as it is by the fluctuating environments of Quaternary time, is less well known. Each author has brought his own special interests and expertise to the synthesis of the region he knows best. To conclude the series, a summary volume by the editors will seek common strands in the regional summaries and erect a national framework within which the more detailed regional accounts will fit.

While no standard approach has been imposed on the contributing authors, care has been taken to be consistent in the use of terminology and the recently-elaborated Quaternary time-scale (Mitchell *et al* 1973) has everywhere been used. Correlation tables are provided for each region, while the maps and diagrams will be found particularly valuable since those in the published literature are so scattered, and generally refer to quite small areas. Above all, these volumes will make the existing literature accessible to the greater number who seek a wider evolutionary understanding of the landforms of the very diverse regions of the British Isles.

Preface

It was in 1685 that a member of the Dublin Philosophical Society made one of the earliest contributions to our understanding of the geomorphology of Ireland; the present volume seeks to summarize our knowledge of that same subject 300 years later. We have not hesitated to indicate where problems still remain, and it must be stressed that while we have necessarily drawn heavily upon the work of others, the conclusions here presented remain our responsibility.

Among those many field-scientists who have lent us their help, especial mention must be made of Francis M. Synge who has been most generous both in providing unpublished data for Chapter 6 and in commenting so constructively upon Chapter 7. We are most grateful to him for giving of his time and for his friendship over many years. The following have kindly permitted reproduction of both published and unpublished material and for this we wish to express our sincere thanks: E.A. Colhoun, J.R. Creighton, R.S. Crofts, M.A. Hannon, A.R. Hill, A.M. McCabe, D.B. Prior, A.G. Smith, W.A. Watts, H.E. Wilson. Professors E.E. Evans, J.P. Haughton, W. Kirk and G.F. Mitchell have in various ways stimulated and supported our studies from the two university centres in Belfast and Dublin, and colleagues in the departments of archaeology, botany, civil engineering, geology and geography, together with officers of the respective Geological Surveys, have also greatly assisted. One of us (N.S.) would like to thank especially A.E.P. Collins, G. Cruikshank, R.E. Glasscock, D.B. Prior, V.B. Proudfoot and L. Symons for companionship in the field, and stimulation in geographical research over a period of some 22 years while he was a member of staff at Queen's University, Belfast. Professor Kenneth Walton of the

University of Aberdeen has also supported the writing of this volume in many ways.

It is pleasant to record our thanks to the cartographers and secretaries who have drawn the figures and typed out the text — Mrs Martha Lyons, Mrs Eileen Russell, Mrs Alison Sandison and Mrs Yvonne Smith.

Finally, we must express our thanks to our respective wives, who have watched us disappear over various horizons to reappear later in need of instant resuscitation! We thank them for their patience and forebearing over many years.

G.L. HERRIES DAVIES   NICHOLAS STEPHENS

Note

Understanding of the chapters that follow will be greatly facilitated by regular reference to the Half-Inch maps of Ireland. The whole island is covered in twenty-five sheets and the sheets carry the Irish National Grid which is used in various places in the text.

1 Geological background

Many a distinguished geologist has chosen to regard Ireland as a mere geological appendage of Britain, and while such a view hardly commends itself to an Irish geologist, there is no escaping the fact that Britain and Ireland do indeed share a common geological heritage. In the far west and northwest, for example, from the crenulated Galway coast to the bold headlands of Donegal, wide areas are underlain by the gnarled schists and gneisses which form Ireland's counterpart to the Moinian and Dalradian rocks of the Scottish Highlands. Similarly, the Ordovician and Silurian rocks of Scotland's Southern Uplands find their Irish equivalent in the Longford-Down axis which strikes southwestwards from the Ards Peninsula for almost 200 km before finally dying out in the ill-drained country around the river Shannon (figs. 1.1 and 1.2). Still farther to the south, Ordovician rocks comparable to those of Wales re-appear on the opposite shores of St George's Channel in counties Wicklow, Wexford and Waterford, there to form an impressive bastion in the angle lying between Ireland's eastern and southern coasts. These Silurian and older rocks are the geological foundation stones upon which Ireland rests. All of them are metamorphosed to a greater or lesser degree, and almost everywhere their southwesterly trending structures reveal the firm imprint of Caledonian orogenesis. One other legacy of that same orogeny is strikingly evident upon any geological map of Ireland — granite. The island contains five major Caledonian granite bodies (the Leinster, Galway, Foxford, Donegal and Newry granites) dating from around 380 million years ago, all of them, save the Galway granite, being elongated along the Caledonian strike. The Leinster granite affords the largest granite exposure to be found within the British Isles; it extends from the sandy southern

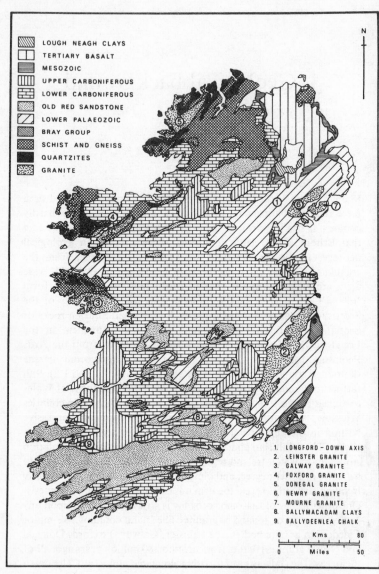

LOUGH NEAGH CLAYS
TERTIARY BASALT
MESOZOIC
UPPER CARBONIFEROUS
LOWER CARBONIFEROUS
OLD RED SANDSTONE
LOWER PALAEOZOIC
BRAY GROUP
SCHIST AND GNEISS
QUARTZITES
GRANITE

N

1. LONGFORD – DOWN AXIS
2. LEINSTER GRANITE
3. GALWAY GRANITE
4. FOXFORD GRANITE
5. DONEGAL GRANITE
6. NEWRY GRANITE
7. MOURNE GRANITE
8. BALLYMACADAM CLAYS
9. BALLYDEENLEA CHALK

Kms 80
Miles 50

Fig. 1.1 Generalized geological structure.

2

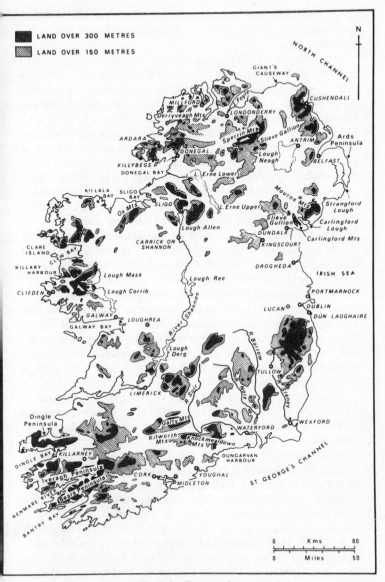

Fig. 1.2 Topography.

3

shores of Dublin Bay southwestwards for 110 km before finally disappearing on the forested western slopes of Brandon Hill in Co. Kilkenny.

At the end of the Caledonian orogeny Ireland again became an area of sedimentation as Devonian and Carboniferous rocks were laid across the contorted Lower Palaeozoic foundations. Eventually these new strata overlapped onto the recently unroofed Leinster and Galway granites, and the sedimentary pile continued to grow until Ireland had received a widespread mantle of those Coal Measure strata which today survive only as scattered outliers. This episode of sedimentary accretion was terminated by the earth movements of the Armorican orogeny, the impact of which was felt in Ireland with an intensity that diminished progressively northwards. It is now usual to recognize three Irish zones of differing Armorican influence (Gill 1962). Firstly, in Munster southward of a line drawn from Dungarvan Harbour to Dingle Bay, the earth movements produced a series of closely-spaced folds which to this day control the entire character of the topography in Waterford, Cork and Kerry. Secondly, to the north of the Dungarvan to Dingle line but southward of a line drawn between Galway and Drogheda, the orogeny has resulted in a more open pattern of folding along posthumous Caledonian lines, thus creating southwestward striking structures in strata of Upper Palaeozoic age. Finally, and to the north of the Galway to Drogheda line, the orogeny produced only mild deformation, and even today the Carboniferous strata are but little disturbed in areas such as counties Leitrim and Fermanagh.

Ireland has an area of 83 000 km² — roughly half the size of England and Wales — and more than 40% of its surface is underlain by Carboniferous sediments. The entire heart of the island is an extensive sheet of Carboniferous strata pierced by only a few scattered inliers; the main road across Ireland from Dublin to Galway — a distance of 212 km — runs over Carboniferous rocks for all but 2 km of its length. Only in peripheral regions, where Ireland's Carboniferous coat has become tattered, do pre-Carboniferous rocks outcrop extensively, and only in the northeast are Mesozoic and Tertiary rocks of any significance. The fact that over the greater part of Ireland the Carboniferous strata are the youngest solid rocks present led an earlier generation of geologists to conclude that the greater part of Ireland had stood above sea-level throughout post-Carboniferous time. But they were wrong. The presence of a large downfaulted outlier of Trias near Kingscourt, Co. Cavan, and of a tiny outcrop of Cretaceous (Senonian) chalk at Ballydeenlea [V952974] 10 km to the north of Killarney (Walsh 1959-60, 1966), both indicate the likelihood of Ireland's Carboniferous rocks having once been buried beneath Mesozoic strata. Denudation has done its work, however, and today it is to the northeast that we must turn in order to find the last shreds of the Mesozoic mantle. There, in eastern Ulster, Triassic sandstones floor such areas as the Lagan valley around Belfast and part of the eastern shores of Lough Foyle, while the Jurassic and Cretaceous strata merely peep out from beneath a protective roof of Tertiary plateau basalts. Indeed, in many places the Jurassic and Cretaceous outcrops have a vertical rather than a horizontal extent and their representation upon even a large-scale geological map perforce has to be somewhat diagrammatic.

The contrast between the startling whiteness of the Cretaceous chalk and the drabness of the overlying black basalts must leave a lasting impression upon any traveller following the Antrim Coast Road, but while the basalts may not be among the most attractive of Ireland's rocks, they do give rise to Ireland's most renowned geological locality — the magnificent Giant's Causeway located on the north Antrim coast. The 100-m high cliffs at the Causeway mark the northern end of a basaltic plateau that covers 3900 km^2 in Antrim and the adjacent counties, a plateau which constitutes by far the largest remnant of the British Tertiary lava fields. The basalts were extruded during Palaeocene time, some 65 million years ago, as the Eurasian and North American plates moved apart to create the Atlantic; they were sub-aerial in origin, the wide extent of individual flows suggesting highly fluid lavas; and they attain a maximum known thickness of 790 m at Langford Lodge to the southwest of Antrim town. The basaltic pile is divisible into three main components. Firstly, there are the Lower basalts which are both the most widespread and the thickest (533 m at Langford Lodge). Secondly, and restricted to northern Co. Antrim, are the Middle basalts which reach a maximum thickness of some 136 m. Finally there are the Upper basalts existing chiefly as scattered residual outliers. During the period covered by the extrusion of the basalts there were occasional spells of quiescence when the youngest lava-flows were weathered under the sub-tropical conditions that then prevailed in Ireland. The result is the red lateritic horizons, each up to 6 m in depth, which are today to be seen in so many basaltic exposures. But one period of quiescence far exceeded all others in its duration — it may have lasted as much as a million years — and deep tropical weathering then produced the well-marked Interbasaltic Horizon that lies between either the Lower and the Upper basalts or, in northern Antrim, between the Lower and the Middle basalts. In a borehole at Washing Bay, on the southwestern shores of Lough Neagh, the horizon was found to be 27 m thick (Wright 1924). It is a striking feature of the Giant's Causeway where the bright red band outcropping half way up the cliffs makes a vivid splash of colour against the sombre hues of the basalts themselves (Cole *et al*. 1912; Eyles 1952).

During and after the extrusion of the basalts, the northeast experienced warping and faulting — some of the faults throw many hundreds of metres — and as a result the base of the basalts varies in altitude from 300 m in the scarps overlooking the North Channel to an estimated ⁻820 m at Washing Bay. The morphology of the area still owed much to this diastrophism and it was as a result of the movements that the basaltic plateau assumed the form of a large tectonic basin. Debris fed into the basin formed a series of soft, pale sands and clays — the Lough Neagh clays — which today underlie some 500 km² around the southern shores of Lough Neagh. The Washing Bay borehole revealed the clays to be more than 350 m thick (Wright 1924) and on pollen evidence they are now ascribed to the period between the later Eocene and the middle Oligocene (Watts 1970).

The effects of the Tertiary igneous activity are by no means confined to the region of the present basaltic outcrop. Roughly contemporary with the extrusion was the cauldron subsidence which admitted the five different granites which today form the Mourne Mountains of Co. Down, while only a few kilometres away, to the west and southwest, further igneous activity yielded the intrusive complexes of Slieve Gullion and the Carlingford Mountains. Much less localised is the great Tertiary dyke-swarm which sweeps across the whole of northern Ireland from Co. Down westwards and northwestwards into Donegal, Sligo and Mayo. Many eastern members of the swarm are finely displayed in coastal exposures between Dundalk Bay and Strangford Lough, while in the west, on the borders of counties Sligo and Mayo, some 35 dykes outcrop on the eastern side of Killala Bay in the 10 km between Inishcrone and

Lenadoon Point. Of comparable age, although different in character, are the dolerite plugs which puncture the surface near Donegal town and in Co. Galway at Doon Hill, near Bunowen, which lies some 10 km southwest of Clifden. Doon Hill rises 60 m above the surrounding schistose lowland and the spheroidal weathering of the plug affords an unexpected contrast to the weathering forms developed in the host rocks. One other intrusive structure deserves mention because of its southerly location and its surprisingly recent age. In 1974 Morris described a dolerite dyke of presumed Tertiary age which traverses the Beara, Iveragh and Dingle peninsulas in counties Cork and Kerry over a distance of 70 km, and later studies in the Dingle peninsula have brought to light a series of other dolerite dykes, for which a dating by the K-Ar method has yielded ages between 42 and 25 million years (Horne and Macintyre 1975). Clearly the area affected by Tertiary volcanism was of far greater extent than had hitherto been suspected.

This Tertiary igneous activity outside the northeast was doubtless accompanied by earth movement of various types. Some of the faulting in counties Galway, Mayo and Donegal, for example, is at least partly Tertiary in age, and the same is true of the faulting at Kingscourt, Co. Cavan. Equally, major faults of Tertiary age are now known to exist in the Irish and Celtic seas, and these seas in all probability occupy rift structures whose foundering was at least partly an event of the last sixty million years (Naylor and Mounteney 1975, *passim*). The sea had certainly entered the Irish Sea basin by the Pliocene because the Pleistocene drifts of Co. Wexford (the so-called 'Manure Gravels') contain ice-dredged Pliocene mollusca of Crag age (McMillan 1964). Since Tertiary deposits are absent from the greater part of Ireland, however, there can only be speculation as to the magnitude of any Tertiary earth movements that may have occurred beyond the limits of the northeastern basaltic province.

Outside the northeast, Ireland contains only one proven Tertiary deposit. At Ballymacadam [S074233] in the Suir valley of Co. Tipperary there lies a small deposit of pipeclay and lignite containing pollens comparable to those found in the Lough Neagh clays. The Ballymacadam deposit would thus seem to be of early or middle Tertiary age and an Upper Eocene date is perhaps the most likely (Watts 1957). In addition to the Ballymacadam deposit, there are a few other sites in southern Ireland where material of presumed Tertiary age occurs. In Co. Kerry, for instance, in the Gweestin valley near Listry Bridge [V866974], there are a number of outliers of an unfossiliferous breccia which are regarded as terrestrial deposits formed by the early Tertiary collapse of the roofs of former limestone caverns (Walsh 1965). Similarly there exist a number of weathering residues which may with confidence be ascribed to the Tertiary. One area where such deposits exist is in Co. Cork near both Midleton and Youghal where Murphy (1966) has employed geophysical methods to locate residues occupying deep solution features within the limestone of the Cork/Castlemartyr syncline. But the best known of the presumed Tertiary residues is the metaliferous 'black mud' which in 1961 was found occupying a fault-guided solution trough developed where the Carboniferous strata meets the Old Red sandstone at Tynagh near Loughrea in Co. Galway. The mud is 45 m deep and it has been worked with great success for silver, lead and zinc.

The post-Carboniferous hiatus in the stratigraphy of the greater part of Ireland may render difficult the interpretation of Tertiary physiographical history, but from the Pleistocene there survives a wealth of geological and geomorphic evidence, and it is proving possible to reconstruct Ireland's glacial history in considerable detail. Save for a few small upland areas in the south and southwest, the whole of Ireland lay beneath the ice during the Munsterian Glaciation (Wolstonian), and during the subsequent Midlandian Glaciation (Devensian) the island was again widely mantled in ice, the chief ice-free region now being a belt of country extending from Co. Wexford to Co. Limerick and separating a Cork-Kerry ice-cap from the main ice-sheet lying farther to the north (see fig. 6.1, p.117). The effects of these glaciations are to be seen in the ice-scoured terrain of Connemara, in the glacially deepened glens of Donegal, in the spectacular glacial drainage channels of Wicklow, and in impressive cirques such as those cut into the eastern face of the Comeragh Mountains of Waterford. Ireland's most notable legacy from the Pleistocene is, nevertheless, her morainic and glacio-fluvial deposits. It is no accident that two Irish words — drumlin and esker — have found their way into international geomorphic terminology, or that it was an Irishman — Charles Smith — who in 1744, when describing the drumlin-swarm of Co. Down, first likened the topography to the surface of a basket of eggs.

The dampness of the Irish climate, and the ill-drained nature of so much of the surface, together give rise to one other characteristic Irish drift deposit — the peat-bog. Peat is estimated to cover almost 15% of Ireland's surface. Raised or high bogs, averaging six m in thickness, predominate in the Midlands, while the rather thinner blanket bogs are located chiefly in the west within the 1000 mm isohyet and in such moist eastern uplands as the Wicklow Mountains and the Antrim Plateau. These great expanses of damp, spongy peat conceal the solid geology and were perhaps one factor that in 1856 caused Sir Roderick Impey Murchison to exclaim: 'I really must declare that the geology of Ireland is the dullest which I am acquainted with in Europe.' But that was an unfair comment. Over the last hundred years many geologists have found satisfaction through a testing of their intellectual mettle against the complexities of Irish geology, and for the geomorphologist, too, Ireland presents a multitude of fascinating problems.

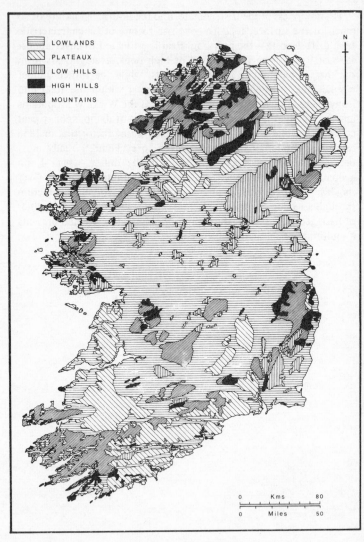

LOWLANDS
PLATEAUX
LOW HILLS
HIGH HILLS
MOUNTAINS

N

0 Kms 80
0 Miles 50

Fig. 1.3   Landform types.

2 Rocks and relief

A comparison of the geological map of Ireland with a map of relief
(see figs. 1.1 and 1.2, pp.2 and 3) reveals a fundamental fact of Irish
geomorphology: lithology and relief stand in the closest possible
relationship. On the one hand, the Carboniferous limestone that
outcrops so widely in the interior of Ireland — an area known as 'the
Midlands' — forms there a gently undulating plain lying almost
everywhere below the 120-m contour, the surface of the plain being
broken only by a few scattered uplands developed upon inliers and
outliers. On the other hand, the outcrops of the pre-Carboniferous
strata, and of the Tertiary igneous rocks, are both largely peripheral
in their distribution, and there, in Ireland's seaward margins, they
form a discontinuous upland rim. True, there are some major breaks
in the rim; an interval of 90 km between the Dublin Mountains and
the Carlingford Mountains allows the Central Lowlands to meet the
Irish Sea on a broad front between Dublin Bay and Dundalk Bay,
while in the west the discontinuous nature of the rim permits the
lowland to finger down to the Atlantic at Galway Bay and Clew Bay.
But it is equally true that Ireland's principal uplands are chiefly
located in peripheral positions and not one of the 45 Irish peaks
exceeding 750 m in height stands more than 56 km distant from
salt-water. There is thus much truth in that old aphorism of the
Victorian textbooks: 'Ireland may be compared to a pie-crust, high
at the edges and sunk in the middle.'

The pie-crust form of Ireland is usually regarded as nothing more
than a reflection of the underlying geology. The Carboniferous
limestone of the Midlands is generally held to be much less resistant
to denudation than are the rocks outcropping along Ireland's
seaward margins, and the pattern of uplands and lowlands is

therefore simply viewed as the work of differential denudation. There can be little doubt that the resistance offered to denudation by the limestone is indeed much less than that offered by such rocks as the Dalradian schists or the Old Red sandstone, and Williams (1963, 1968) estimated that the limestone of the Fergus River basin in Co. Clare is undergoing solution at a rate equivalent to the removal of 51 mm from the surface every thousand years. But explanation of the broad pattern of Ireland's topography solely in terms of differential denudation perhaps begs one important question. Since the upland rim is developed in pre-Carboniferous rocks almost everywhere except in the northeast, then why in the Midlands should those same pre-Carboniferous strata mostly lie far below sea-level, deeply buried beneath a heavy Carboniferous overburden? Has the rim experienced differential uplift as compared with the Midlands, and if so, was the movement a sufficiently recent event to offer at least a partial explanation for the elevated character of Ireland's coastal regions? This question will be examined more closely in Chapter 4.

Ireland's mountains, whether in the coastal rim or further inland, are typically arranged in small, compact units separated from each other by lowlands lying within 100 m of sea-level. As a result, Ireland lacks extensive areas of continuous upland such as those which characterise Scotland, Wales, or Northern England. Many of the Irish mountain ranges — the Knockmealdowns, for instance, or the Ox Mountains — are only one mountain wide, and what Linton (1964a) termed the 'fragmentation' of the Irish uplands is particularly well displayed in the west in the country around Killary Harbour. There, in an area of less than 800 km², the mountains are arranged into ten relatively compact, steep-sided units, all but two of them rising to over 500 m, and each separated from its neighbours by valleys and lowlands lying less than 100 m above sea-level (see fig. 3.7, p.47).

In many cases the scale of the Irish uplands is obviously determined by the size of relatively small but resistant inliers rising through the much weaker limestone. In such circumstances extensive uplands are hardly to be expected. In the coastal areas, on the other hand, tough pre-Carboniferous rocks outcrop widely, and the absence of broad highland areas is a puzzling anomaly. The Dalradian rocks of Ulster, for example, might reasonably have been expected to reproduce the kind of terrain which they form in the Scottish Highlands, and similarly it might have been anticipated that the rocks of the Longford-Down axis — the southwestward prolonga-tion of the Lower Palaeozoic country of southern Scotland — would form an Irish counterpart to Scotland's Southern Uplands. In neither of these regions — nor elsewhere — do Ireland's ancient rocks reproduce that type of extensive upland topography to which they give rise in other parts of the British Isles. Ireland's geological foundations, it seems, are in a very much more advanced stage of denudation than are their British counterparts. Linton (1964a) sugges-ted that this fact might perhaps be explained in terms of climatic geomorphology. Could it be that the dampness of the Irish climate allows base-levelling to proceed more rapidly in Ireland than else-where within the British Isles? Because of their more maritime location, counties Donegal and Down, for example, might be regarded as having already attained a stage that will not be achieved in Scotland's Highlands and Southern Uplands until perhaps several million years hence. Interesting though this idea may be, doubts must linger. Is the Irish climate really sufficiently different from that of Britain to place Ireland in a separate morphoclimatic region? If dampness be the crucial factor leading to accelerated denudation, then it is worth noting that no region of Ireland has precipitation values as high as those of the Western Highlands of Scotland, and while the number of rain-days per year in Ireland is high (200 per annum almost everywhere), equally high figures are recorded over most of Highland Britain. For the moment this question of the relative efficiency of denudation in Ireland *vis-à-vis* that in Britain must remain an open question, but the fragmentation of Ireland's uplands is a problem that will again engage our attention in chapter 4.

The Carboniferous limestone must certainly rank as one of Ireland's least resistant rocks, but at the other end of the spectrum which of Ireland's major rock types is in the geomorphic sense qualified to be regarded as the hardest? There are perhaps two chief contenders for the title: the quartzites and the Old Red sandstone. Quartzites are widely scattered amongst the ancient schists of Galway, Mayo and Donegal and in the equally ancient rocks of the Bray Group in Co. Wicklow (see fig. 1.1, p.2). Rarely are the quartzite outcrops of any very great extent, but wherever they occur they give rise to a topography of hills and mountains. Although quartzite underlies only a small fraction of Ireland's surface, it forms no less than 12% of the peaks that top the 600-m contour, many of the quartzite mountains displaying the conical form so characteristic of that particular rock-type. Among such spectacular conical peaks, the two most familiar are probably Errigal in northwest Donegal and Croagh Patrick which sits majestically on the southern side of the drumlin-studded waters of Clew Bay. But perhaps a third quartzite peak merits mention because it must be among the first mountains seen by many a visitor to Ireland's shores: the Great Sugar Loaf of Co. Wicklow which rises on the southern horizon as the boat from Holyhead noses her way into Dun Laoghaire Harbour.

The quartzites outcrop chiefly in the northern half of Ireland; the domain of the resistant Old Red sandstone is confined principally to the south. There the various sandstones, mudstones, and slates that comprise the Old Red sandstone series almost invariably support a topography of hills and mountains, and the juxtaposition of the Old Red sandstone and the Carboniferous rocks commonly results in a very abrupt change in the character of the terrain. Here, however, a caveat would perhaps not be out of place. In many localities the Old Red sandstone/Carboniferous junction is obscured by drift and the officers of the Geological Survey, already familiar with the principles of feature mapping, may in some cases merely have drawn their geological boundaries along the topographical break of slope. The geomorphologist who returns to such areas and then uses the geological map to demonstrate the existence of a close relationship between topography and lithology may unwittingly be doing nothing more than completing the second half of a cyclical argument. But about the mountain-forming potential of the Old Red sandstone there need be no doubt. Although the Old Red underlies less than 10% of Ireland's surface, it forms almost 50% of the 190 or so Irish peaks that rise over 600 m, and among the sandstone mountains is numbered Carrauntoohil, Co. Kerry, the highest peak in Ireland.

Of the other Irish rocks, only two need be singled out for special mention: the Cretaceous chalk and the granites. The chalk can be dismissed very briefly because of its limited outcrop, and all that needs to be noted is that the Irish chalk is much harder than its English counterpart. For long this hardness was believed to be a result of a 'baking effect' of the overlying basalts, but it is now more realistically attributed to the chalk's pore-spaces being filled with fine-grained secondary calcite (Hancock 1961, 1963; Manning *et al*. 1970, pp. 75-5). The Irish granites, on the other hand, are a much more equivocal type of rock; in some places they give rise to mountains while in other localities they underlie extensive lowlands. The Mourne pluton, for instance, forms the bold massif of the Mourne Mountains, while the outcrop of the Galway granite yields only the very much more subdued topography of the Connemara Lowland. Even within the outcrop of a single granite body there are often to be found striking topographical contrasts. Thus, the main mass of the Donegal granite forms the Derryveagh Mountains while the granite's southwestward extension — the so-called Ardara diapir — forms a lowland which is overlooked by a ring of hills developed upon the adjacent schists of the aureole. Even more striking is the case of the Leinster granite. Throughout much of its length this granite forms the crest-line of the Dublin and Wicklow Mountains, including therein the great dome of Lugnaquillia (935 m), the highest peak in Leinster. Southwestward of Lugnaquillia, however, the same granite underlies the Tullow Lowland of Co. Carlow. This variation in the topographical expression of the Irish granites may result from petrological differences within each granite, but in the present state of our knowledge there is little further elucidation that can be offered upon this subject.

The consideration of many other facets of the structural geomorphology of Ireland may conveniently be deferred until chapter 3, but there is one final topic that deserves a brief mention. As we have seen already, many of the structural lineaments of Britain are projected westwards into Ireland, and among these are those three major elements in the geology of Scotland; the Southern Uplands Fault, the Highland Boundary Fault, and the Great Glen Fault. In recent years the ingenuity of many a geologist has been deployed in trying to trace the precise course of these three faults across Northern Ireland. The Southern Uplands Fault is believed to enter Ireland through Belfast Lough and to run thence along the northern margin of the Longford—Down axis to near Carrick-on-Shannon where the fault disappears beneath the Carboniferous limestone of the Midlands. Farther to the northwest the Highland Boundary Fault is usually interpreted as striking southwestwards from Cushendall, Co. Antrim, and running thence along the southeastern flanks of the Sperrin and Ox mountains into Clew Bay before finally disappearing into the Atlantic off the cliffed western coast of Clare Island, Co. Mayo. The Irish course of the Great Glen Fault remains rather more speculative, but it is commonly equated with the Leannan Fault which traverses Co. Donegal along a line from Millford to Killybegs. But the very fact that there is so much uncertainty about the Irish course of all these faults should convey its own implicit message: nowhere in Ireland are these tectonic structures very clearly marked and none of them possesses that profound topographical significance which is theirs in Scotland. In the Giant's Causeway, Co. Antrim may possess the analogue for Scotland's Isle of Staffa, but neither Scotland's Glen More not its Midland Valley find any counterpart in the geomorphology of Ireland.

3 Geomorphic regions

Scotland and Wales are both divisible into a few broad physiographic units; the same can hardly be said of Ireland. There, nature has produced a terrain which is full of finicky detail rather than being rough-hewn; a landscape more akin to an intricate etching than to a crude casting. Almost every Irish county contains considerable topographical diversity, and the fragmentation of the Irish uplands results in such a complex interdigitation of lowland and highland as to render impossible any generalisation of the topography into a few major regions. For our present purposes, therefore, Ireland has been divided into the nineteen morphological regions represented in fig. 3.1, each of the regions possessing a unity based upon a combination of lithological and structural factors.

The Central Lowland (1)

This is Ireland's heart — a lake-studded lowland occupying more than one third of the island (fig. 3.1). Only locally does the surface of the lowland rise above 120 m and large areas lie below 60 m in the Shannon basin, in the vicinity of loughs Mask and Corrib, and around the head of Galway Bay. The lowland is developed chiefly upon gently folded Lower Carboniferous strata, although in the northeast, northward of the river Boyne, the lowland does pass onto the Silurian rocks of the Longford — Down axis. Elsewhere, however, the limits of the lowland are broadly coincident with the limits of the limestone outcrop, although that term 'limestone outcrop' perhaps stands in need of some amplification. It has long been customary to employ a tripartite division of the Irish limestones: Lower limestone, Middle or Calp limestone, and Upper limestone,

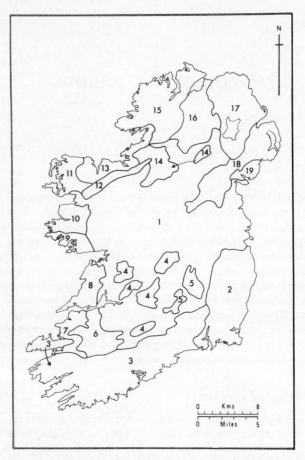

Fig. 3.1 Geomorphic regions (numbers refer to headings in text).

but it must be emphasised that this division is misleading in the sense that many of the beds mapped as limestone really possess a character which is arenaceous or argillaceous rather than calcareous. This is particularly true of the strata which have been mapped as Middle or Calp limestone, and the extent of true limestones over the Central Lowland is certainly far less than the geological map might suggest.

The gently undulating surface of the lowland is broken by minor hills and plateaux developed upon four types of structure. Firstly,

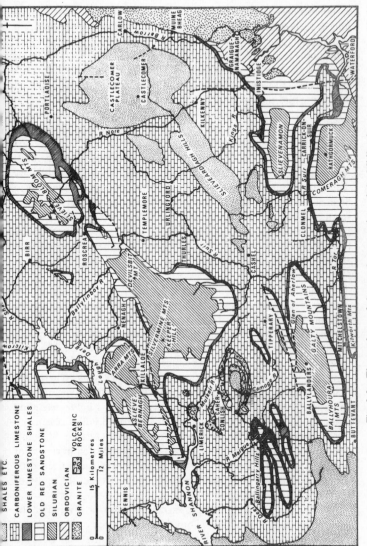

Fig. 3.2   The geology of the southern portion of the Midlands.

SHALES ETC.

CARBONIFEROUS LIMESTONE

LOWER LIMESTONE SHALES

OLD RED SANDSTONE

SILURIAN

ORDOVICIAN

GRANITE   VOLCANIC ROCKS

0 ___ 15 Kilometres
0 ___ 12 Miles

PORTLAOISE

CARLOW

MUINE BHEAG

R. Barrow

CASTLECOMER PLATEAU

CASTLECOMER

GRAIGUE NAMANAGH

SLIEVE BLOOM MTS

BIRR

R. Nore

KILKENNY

R. Nore

Kings R.

INISTIOGE

WATERFORD

R. Suir

R. Brosna

Ballyfinboy R.

ROSCREA

TEMPLEMORE

URLINGFORD

SLIEVEARDAGH HILLS

SLIEVENAMON

CARRICK-ON-SUIR

RATHGORMUCK

DEVILSBIT MT.

THURLES

R. Suir

CASHEL

CLONMEL

R. Suir

COMERAGH MTS

NENAGH

SILVERMINE MTS

KEEPER HILL

R. Suir

TIPPERARY

Glen of Aherlow

GALTY MOUNTAINS

KILLALOE

Lough Derg

ARRA MTS

Mulkear R.

Camogg R.

BALLYLANDERS

BALLYHOURA MTS

MITCHELSTOWN

Kilworth Mts

ASLIEVE BERNAGH

LIMERICK

CONLISHEEN

R. Maigue

Ballingarry Hills

BUTTEVANT

ENNIS

RIVER SHANNON

R. Deel

the limestone is pierced by many small inliers which give rise to such hills as the Chair of Kildare (sediments and volcanics of Ordovician and Silurian age together with Old Red sandstone), the Slieve Bawn ridge to the north of Lough Ree in Co. Roscommon (Silurian strata, Old Red sandstone, and basal Carboniferous sediments), and the Ballingarry Hills of Co. Limerick (volcanic agglomerates, Old Red sandstone and the basal Carboniferous beds).

Secondly, the Carboniferous rocks of the lowland contain much reef-limestone which possesses a lithology very different from that of the surrounding strata. Many of the reefs were originally mounds upon the floor of the Carboniferous sea, and these features have now been exhumed to form prominent reef-knolls. The knolls are commonly circular in plan and their flattish tops may stand up to 60 m above the level of the surrounding plain. Such knolls are widely scattered, but they are particularly well developed to the north of Dublin in an arc extending from Lucan to the sea near Portmarnock (Nevill 1958). Feltrim Hill [0202444], one member of this arcuate chain, has served as a type-example of a reef-knoll for generations of Dublin students.

Thirdly, the limestones in a few places contain contemporaneous igneous rocks which commonly give rise to small ranges of hills. One example lies 6 km north of Daingean, Co. Offaly, where a Carboniferous volcanic neck forms the bold eminence of Croghan Hill [N481331] which rises 120 m above the adjacent lowlands. The most extensive outcrop of the Carboniferous volcanic beds is in Co. Limerick in an area southeastward of Limerick City. There, the igneous rocks form a broad east-to-west trending syncline (the Limerick Volcanic basin) some 20 km in length, its limits extending from Caherconlish in the north, southeastwards to Pallas Grean and then westwards via Herbertstown and Lough Gur, to Shehan's Cross. The structure of the syncline is clearly reflected in ranges of hills possessed of steep outward-facing escarpments and backed by gentle dip slopes that lead down to the Camoge River at the centre of the basin (Coudé 1973).

Finally, the surface of the Central Lowland is diversified by features developed upon small Upper Carboniferous outliers such as those which form the Summer Hill ridge and the Hill of Tara [N919597] in Co. Meath, or the uplands to the west of Kiltamagh in Co. Mayo.

Here must be mentioned one minor feature that breaks the surface of the Central Lowland but which falls into none of the four categories just discussed — the Rock of Cashel, Co. Tipperary. The Rock, so familiar to road travellers between Dublin and Cork, and known to a wider international public through its appearance upon Irish postage stamps, is merely a small, flat-crested limestone anticline which has been left isolated by the denudation of the surrounding strata.

The bold outcrop of the Rock of Cashel is a somewhat unusual feature of the Midland landscape because over most of the Central Lowland the solid-rock morphology is obscured by thick deposits of drift, and such landforms as drumlins, kames, eskers, and peat-bogs provide the lowland with its micro-features. But in many areas to the west of the Shannon the drift is thinner and the limestone locally rises to the surface to form pavements that are diversified by a variety of solution phenomena. Caves also exist, as for example south of Tuam, Co. Galway (Drew 1973), and on the Aille River to the southeast of Westport, Co. Mayo (Coleman 1950, 1965), but the lowland's two most famed areas of karstic phenomena are the Cong region lying between loughs Mask and Corrib (Coleman 1955, 1965) and the drumlin-covered Gort Lowland located to the southeast of Galway Bay (Coleman 1965). The former of these two areas contains an interesting freak — a 'dry canal' which was foolishly built through the fissured limestone between loughs Mask and Corrib and which, despite the best endeavours of the nineteenth-century engineers, has never been persuaded to hold water sufficient to float a barge between the two loughs. The second of the two localities — the Gort Lowland — displays excellent examples of another, but more natural hydrological phenomenon — turloughs. The term comes from the Irish *Tuar Loch* meaning literally 'a dry lake'; it is applied to those hollows which contain ephemeral lakes, and more especially to those hollows which hold water only during the winter months (Williams 1970). Of course, not all the lakes on the Gort Lowland qualify to be regarded as turloughs and it should be noted that far from occupying enclosed limestone depressions, many of the multitude of lakes upon the lowland lie in hollows in the local veneer of glacial drift.

In the eastern portion of the Central Lowland caves and sink-holes are to be found in the limestone that outcrops amidst the drumlins of Co. Monaghan near Carrickmacross (Coleman 1952, 1965), while in the south the large Mitchelstown cave systems seam the limestone

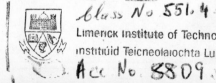

that intervenes between the Old Red sandstone of the Galty and Kilworth mountains (Hill *et al.* 1909; Coleman 1965). In general, however, solution features are rare to the east of the Shannon, although geophysical exploration in the east has revealed some interesting karstic phenomena buried beneath the drifts of counties Meath, Westmeath and Offaly. There, more than twenty intense negative Bouguer anomalies have been located along a southwesterly trending line some 115 km in length, and these are best interpreted as large caverns developed along a line of structural weakness (perhaps a major fault) and filled either with water or with unconsolidated debris (Murphy 1962).

Solution of the limestone must certainly have played a part in the development of some of the many lake basins which are scattered over the surface of the lowland, but glacial erosion has probably been a more important formative process in the case of the larger lakes (Charlesworth 1963c; Williams 1970). Of these larger lakes, Lower Lough Erne is the deepest (64 m), but loughs Mask, Corrib, Ree and Derg all attain depths in excess of 30 m. The floors of loughs Mask, Corrib, Derg and Lower Lough Erne all lie below sea-level, and both Lough Mask and Lough Corrib have bottoms more than 36 m below Ordnance Datum. On the other hand, a large number of the smaller lakes merely occupy shallow hollows in the peat mantle, or else they are dammed up by nothing more substantial than accumulations of glacial debris. Upper Lough Erne, for example, that remarkable interlacement of land and water in Co. Fermanagh, is simply a result of the flooding of the hollows in a part of the great drumlin swarm of Northern Ireland.

Both Lough Ree and Lough Derg lie on the course of the River Shannon which is by far the most important of the rivers that drain across the Central Lowland. Indeed, the Shannon catchment encompasses about one-fifth of the surface of Ireland. The river rises in the Cuilcagh Mountains of Co. Cavan, only 40 km from the shores of Sligo Bay, but its waters flow southwards to reach the sea below Limerick, and its course of some 280 km makes the Shannon the longest river in the British Isles. For most of its course the river has only the gentlest of gradients and it falls a mere 15.5 m in the 220 km between loughs Allen and Derg (Kilroe 1907b). It need, therefore, evoke no surprise that flooding has always been a serious problem along the course of the river northward of Lough Derg, the marshy areas on either side of the river being known as 'callows'. Between

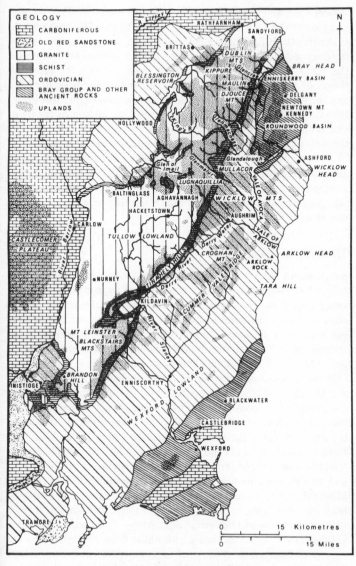

Fig. 3.3 Southeastern Leinster.

Lough Derg and Limerick, however, the gradient is much steeper and the river falls a distance of 30 m in only 30 km. This sharp break in the river's profile is as yet inadequately explained and the same is unfortunately true of another and even more remarkable feature of the Shannon in the vicinity of Lough Derg. Around the southern shores of the lough there lies a large upland — the Slieve Bernagh and the Arra Mountains — developed upon a major inlier formed in Silurian strata and the Old Red sandstone, but instead of taking a simple course over the limestone lowlands lying either to the east or the west of the inlier the Shannon flows across the inlier by means of a deep gorge developed around Killaloe. For long the Killaloe gorge was regarded as a result of the Shannon's superimposition, but a recent study (Farrington 1968) has cast grave doubt upon such an interpretation and the issue must still be regarded as a very open one (see chapter 5).

The Leinster Axis (2)

Southeastern Ireland is an area of marked Caledonian structure and its southwestward striking grain is picked out by many a subsequent valley (fig. 3.3). The region's backbone is the Leinster mountain chain which extends southwestwards from the shores of Dublin Bay for a distance of more than 110 km. The chain includes both Ireland's largest area of continuous upland standing at an altitude of over 300 m, and Ireland's highest granite mountain — Lugnaquillia (935 m). Geologically the region is dominated by the Caledonian granite of the Leinster batholith which is now known to be a compound body consisting of five dome-shaped units emplaced more or less synchronously. Brindley (1973) has given the five units the following titles: the Northern Unit, the Upper Liffey Unit, the Lugnaquillia Unit, the Tullow Lowlands Unit, and the Blackstairs Unit (fig. 3.4). In the north, in the southern suburbs of Dublin City, and in the west, in Co. Carlow, the limestone of the Central Lowland overlaps onto the granite, but elsewhere the granite body is enclosed within a broad outcrop of Lower Palaeozoic slates and grits which are chiefly of Ordovician age. Along their contact with the granite these Lower Palaeozoic rocks have been converted into an aureole of mica-schist which is of some morphological significance because the schists, together with the indurated marginal granite, offer more resistance to denudation than do either the non-schistose

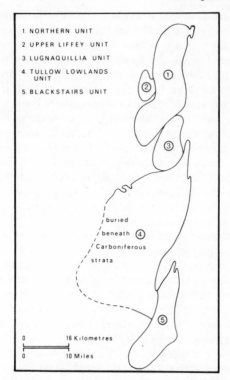

1. NORTHERN UNIT
2. UPPER LIFFEY UNIT
3. LUGNAQUILLIA UNIT
4. TULLOW LOWLANDS UNIT
5. BLACKSTAIRS UNIT

buried beneath Carboniferous strata

0   16 Kilometres

0   10 Miles

Fig. 3.4   The units of the Leinster Batholith (after Brindley 1973).

Lower Palaeozoics or the main body of the batholith. Over much of its length the roof of the batholith dips gently to the southeast, whereas to the northwest it plunges steeply, and this asymmetry results in the outcrop of the schist being widest on the southeastern side of the granite, although even there the aureole is rarely more than 3 km wide.

In central Wicklow, where the granite reaches its highest elevation, there are large numbers of schist bodies set into the granite of the Lugnaquillia Unit. Most of these bodies appear to be fragments of former roof-pendants, but on the dome-shaped summit of Lugnaquillia itself there lies a large cake of schist which is a surviving fragment of the actual roof of the batholith. Farther to the north, overlooking Dublin, the summits of some of the Dublin

Mountains are developed in a type of granite which is characteristic of the batholith's marginal zone (Brindley 1954), so there too the mountain tops evidently lie close to the former roof of the granite body. Thus, although the flanks of the northern portion of the Leinster granite have been deeply dissected, it seems that relatively little granite has been lost from the crest of the batholith. The possible significance of this fact will be explained in the next chapter.

Although the Lower Palaeozoic rocks encircling the Leinster batholith are predominantly slates and grits, they are associated with a variety of contemporaneous igneous rocks. These igneous rocks — ashes, tuffs, lavas, sills, and dykes — are distributed widely, but they are particularly important in two tracts of country, one following the Caledonian strike southwestwards from Wicklow Head into northern Co. Waterford, and the other taking a parallel course from Arklow Head to the south coast in the vicinity of Tramore. The only other rocks deserving of mention within the Leinster Axis are the somewhat enigmatic strata of the Bray Group which are usually regarded as being of either Cambrian or pre-Cambrian age. They outcrop widely in northeastern Co. Wicklow and southeastern Co. Wexford and they consist chiefly of purple grits and slates. But within the rocks of the group — and this is the reason for their receiving a particular mention here — there are beds of quartzite which form many prominent landmarks.

Such, then, is the geology of the Leinster Axis, and we can turn now to the geomorphology of the area, taking first the Leinster mountain chain which constitutes the region's backbone. The chain falls readily into two portions. Firstly, there is a broad northern portion — the Dublin and Wicklow mountains — lying between Sandyford, Co. Dublin, and Lugnaquillia. Secondly, there is a much narrower portion extending southwestwards from Lugnaquillia to the incised meanders of the Nore around Inistioge, Co. Kilkenny, the mountains being flanked to the northwest by the extensive Tullow Lowland.

The core of the northern portion of the chain is the Leinster batholith, and the granite forms the peat-mantled array of mountains and upland basins which is displayed so superbly to the traveller on the Military Road as he wends his way southward from Rathfarnham to Aghavannagh. The granite terrain mostly stands at over 360 m, and the highest mountains lie along the median-line of the granite outcrop where more than a dozen broad, convex summits rise to over 600 m. The subdued character of the squat granite domes was once attributed to the influence of an overriding ice-sheet, but it is now recognised that sheet jointing within the granite has probably played a major role in giving the topography its flowing form. Tors are few, and the 'sharpest' features developed in the granite are a few cirques that locally give rise to a mild form of biscuit-board topography as, for example, where the cirques containing the two Bray loughs fret the eastern slopes of Kippure.

On either side of the batholith the tough schist and indurated granite give rise to features lower, but rather sharper, than those developed upon the main body of the granite. Many of the rivers draining off the granite show a constricted valley section where they cross the aureole (Farrington 1927). Since the aureole is at its widest to the southeast of the batholith, it is on that side that the influence of the resistant schist is best displayed. The ridge of hills to the north of the Enniskerry basin, together with the peaks of Maulin, Djouce and Mullacor, are just a few examples of the relatively bold features that are associated with the southeastern aureole, while the constriction of a valley in the schist is exemplified in Glencree where the mass of Knockree [0185159] protrudes into the glen and reduces it to a fraction of the width it possesses on the granite further upstream. On the opposite, northwestern side of the batholith, the aureole is narrower, but it still exerts its influence upon the topography. The aureole rocks cause the constriction in the valley of the Ballydonnell Brook [0060120], a tributary of the Liffey, and they underlie both the narrow ridge which protrudes into the Blessington Reservoir to the northeast of Hollywood and the bold line of hills about Baltinglass.

Back on the southeastern side of the batholith is one other feature related to the relative resistance of the schist as compared with that of the granite. The southeastern flanks of the Wicklow Mountains are trenched by the deep, U-shaped forms of the picturesque Wicklow glens. Three of the glens — Glenmacnass, Glendasan, and Glendalough — are developed chiefly in the schist and the Ordovician strata, and they are each terminated very abruptly by trough-end features located at about the granite/schist junction. The streams draining through the glens thus possess courses that are readily divisible into three reaches: firstly, a gently graded upland reach in a granite basin lying at a height of over 300 m; secondly, a short but spectacular reach where the streams cascade down the walls of the trough-end features; and finally, another gently graded reach where the streams traverse the drift infill on the floor of the glens at a height of only 180 m or so. Farrington (1927) ascribed both the excavation of the glens and the production of the trough-end features to differential glacial erosion. He pictured the pre-glacial valleys — like so many modern Wicklow valleys — as having been constricted in the schist, so that the Pleistocene glaciers streaming down off the granite uplands were thrown into pressure-ridges and thickened as they crossed the aureole. This thickening of the ice, he suggested, resulted in accelerated glacial erosion of the schist leading eventually to the overdeepening of the glens. Thus, paradoxically, he related the development of trough-end features in the schist to the resistance which the same rock had been able to offer to the fluvial processes operating during pre-glacial times.

Ingenious though it is, this explanation seems hardly adequate to account for the form of Glenmalur, which is in many respects the most impressive of the Wicklow glens, because there the trough-end feature lies in the granite 5 km upstream of the granite/schist junction. Perhaps the glens are really composite features, partly the work of glacial processes, and partly the work of vigorous fluvial erosion following a general rejuvenation of drainage. The proximity of the cascades to the granite/schist junction in Glenmacnass, Glendasan and Glendalough may be the result of the recession of rejuvenation heads having been retarded by the resistance offered upon the schist. In passing it may be noted that there is no counterpart to the Wicklow glens on the western side of the chain. Instead the granite there has been opened out to form the three very broad basins of the Liffey, the Kings River, and the Glen of Imail, the floors of which all lie within 210 m of sea-level.

Outside the aureole zone, the foothills of the northern portion of the Leinster chain are developed in the Lower Palaeozoic rocks and the members of the Bray Group. Together these rocks form a series of upland basins (one of the most interesting of these is the Roundwood Basin of northeastern Wicklow) and some rather subdued hills, most of which do not exceed 390 m in height. The beds of quartzite in the Bray Group, and the igneous rocks amongst the Ordovician strata, nevertheless give rise to some bolder features. The quartzite forms Bray head, the Great and Little Sugar Loafs, and Carrick Mountain, while the Ordovician volcanics form the uplands flanking the mine-scarred Vale of Avoca.

Two other features in the northern portion of the Leinster chain merit brief mention. Firstly, in the foothills to the northeast of the chain there occurs a series of most impressive glacial drainage channels, The Scalp [0214205] and The Glen of the Downs [0260110] being excellent examples. Secondly, in eastern Co. Wicklow there is a sinuous escarpment separating the narrow, drift-covered coastal-plain from the hills and upland basins farther inland. This feature — it may be termed the 'East Wicklow Escarpment' — is particularly well seen between Bray Head and Wicklow Head, in which district it lies to the west of Delgany, Newtown Mt Kennedy and Ashford. Southward of Wicklow Head the escarpment is somewhat broken, but it can be traced at least as far as Shelton Abbey in the Vale of Arklow. The escarpment is developed in strata of both Bray Group and Ordovician age, and although locally it may owe its form to the presence of the Bray Group quartzites, the existence of the feature as a whole cannot be explained simply in lithological terms. Perhaps the escarpment is a fault-line scarp which has retreated westwards following the Tertiary earthmovements in the Irish Sea basin.

Turning now to the southern portion of the Leinster chain — that lying southwestward of Lugnaquillia — we find that the chain has lost much of its grandeur because the granite has ceased to be a mountain-forming rock. Instead, it underlies the Tullow Lowland, and the line of the Leinster chain is continued by a comparatively narrow, disjointed belt of hills and mountains developed upon rocks lying along the batholith's southeastern margin. Between Aughrim and the River Slaney these uplands are arranged in two parallel ridges separated by the long strike valley which runs for 32 km from Aughrim to the Slaney at Kildavin. The hills to the northwest of the valley (here termed the 'Tinahely Hills') are developed upon the marginal granite and the schist, and the southeastern ridge (the 'Cummer Vale Ridge'), which includes the prominent Croghan Mountain, is formed chiefly in the Ordovician volcanics. One puzzling point to note about the Tinahely Hills is that they do not form a water-parting. No less than five streams flow southeastwards off the Tullow Lowland and through the hills in order to join either the Derry Water or the Derry River in the strike valley.

Southwestward of the incised and supposedly superimposed course of the Slaney, only one ridge is present — the narrow, tor-capped ridge of the Blackstairs Mountains which rise to 795 m in Mount Leinster. These mountains are developed upon the indurated rocks of the granite/schist junction along the southeastern side of the Blackstairs Unit of the batholith, and the marked asymmetry of the ridge is evidently related to the batholith's structure. The low southeastward dip of the granite/schist junction is reflected in the comparatively gentle and convex southeastern flanks of the mountains, whereas on the opposite side of the ridge the slopes are concave and extremely steep, where denudation has eaten out the core of this particular granite unit along the line of the Barrow valley. Across the Barrow (another supposedly superimposed river), the southwestern extremity of the main mass of the Leinster granite forms Brandon Hill which shows the same asymmetry but in reverse; its eastern, concave face rises steeply from the Barrow while its more gentle, westward slopes overlook the Nore valley and seemingly reflect the gentle southwestward dip of the roof of the Blackstairs Unit. Beyond the Nore we need not stray at present; there the Leinster chain dies out in a belt of hills as the Caledonian trend of Eastern Ireland gives way to the Armorican structures of the South.

To the north of the Blackstairs Mountains and to the northwest of the Tinahely Hills there lies the broad expanse of the Tullow Lowland developed upon the granite of the Tullow Lowlands Unit. The surface of the lowland varies in height from 60 m to 240 m and it consists of a series of very shallow, ill-drained basins, many of which contain thick accumulations of peat. Excavations for land-drainage schemes repeatedly reveal vast numbers of granite boulders having the form of core-stones left by the rotting of the parent rock. A few gentle hills diversify the landscape as, for example, around Nurney, but the most prominent features are some small, inselberg-like hills which rise abruptly out of the surrounding lowland. One of the best instances of this type of feature is the quartzose Eagle Hill [S975788] near Hacketstown; the hill is well worth climbing for the splendid view. The southeastern and southern skyline is formed by the Tinahely Hills and the Blackstairs Mountains; in the west there looms the tableland of the Castlecomer Plateau; to the north lies the line of hills developed upon the northern aureole of the granite around Baltinglass; and to the northeast the upland ring is completed by the bold granite mass of Lugnaquillia. The *vista* poses an obvious question: why should the granite here have been reduced to a hill-girt lowland when immediately to the northeast the granite rises to form the impressive upland that is the Wicklow Mountains? If it be accepted that the Leinster granite as a whole is relatively weak rock, then that question might be answered by suggesting that until recently the granite from Lugnaquillia northwards was protected from denudation by the survival of its schist roof, whereas farther to the south the granite was unroofed much earlier, thus allowing denudation ample time in which to produce the Tullow Lowland. Certainly, as we have seen, there is clear evidence that even today denudation in the Wicklow and Dublin mountains is still working upon the granite at a level but little below that of its original roof. Equally, the overlap of the Carboniferous limestone onto the granite of the Tullow Lowland along a wide front just to the east of the River Barrow leaves no doubt that the granite of the Tullow Lowland was exposed as far back as the Upper Palaeozoic (Davies 1960a). Material evidence therefore exists to support the explanation just offered, but at the same time it should be noted that the limits of the Tullow Lowland are closely coincident with those of the granite of the Tullow Lowlands Unit. There is thus the possibility that some petrological variation within the granite units has resulted in the Tullow Lowlands Unit proving particularly susceptible to denudation.

Southeastward of the Cummer Vale Ridge and the Blackstairs Mountains there lies the final portion of the Leinster Axis to command our attention — the Wexford Lowland. The undulating surface of the lowland occupies the whole corner of Ireland lying to the southeast of a line drawn from Arklow to the shores of the Celtic Sea at Tramore. Its surface is widely carpeted with glacial drifts (especially noteworthy is the superb morainic topography of the Screen Hills between Blackwater and Castlebridge) and its form is diversified by a number of steep-sided hills which have been interpreted either as inselbergs or as the cliffed islands of some former transgressive sea. A few of these hills, such as Forth Mountain near Wexford town, are developed upon the Bray Group quartzites, but most of them are associated with the belt of Ordovician igneous rocks that strikes southwestward across the region from Arklow Head. It is these rocks that form features such as Arklow Rock, Tara Hill, and the upland country around Enniscorthy, including there that eminence famed in Irish military history — Vinegar Hill.

The Ridge and Valley Province (3)

The character of this region is determined entirely by the presence of the well-developed Armorican fold-system of southern Ireland. Rocks of three basic types are involved in the folding (fig. 3.5). Firstly, there is the Old Red sandstone, here somewhat unfortunately titled because its strata include sandstones, mudstones, slates, and conglomerates ranging in colour from reddish-brown and purple to green and grey. Over the greater part of Ireland the Old Red sandstone is never more than 1200 m thick, but here in the south the series thickens rapidly to attain in Kerry a maximum known Irish thickness of more than 6700 m. Secondly, and resting upon the Old Red sandstone, there are Carboniferous strata consisting of thin basal beds of the Lower Limestone shales followed by a thick sequence of Carboniferous limestone. To the south of a line drawn from Cork Harbour to Kenmare, however, there is a change in the character of the Carboniferous sediments, and instead of being mantled with Lower Limestone shales and the limestone, the Old Red sandstone is overlain by a thick series of arenaceous and argillaceous rocks which constitute the third and final rock category present in the region. These rocks of the extreme south appear upon the maps of the Geological Survey as the Carboniferous slate and the Coomhola grits, but they are now known more simply as the Cork Beds. If the limestone ever existed southward of the Cork Harbour to Kenmare line, then it must have fallen victim to intra-Carboniferous denudation because in the Bandon and Owenboy valleys, and at the Old Head of Kinsale, small outliers of Namurian strata rest directly upon the Cork Beds.

Within all these rocks the Armorican folds trend from east to west in Co. Waterford and eastern Co. Cork, but farther to the west the structures swing towards the west-south-west before fingering out into the Atlantic to form the famed ria coastline of the southwest. In the past the folds were commonly regarded as simple symmetrical structures, but recent research suggests that at least some of the folds are markedly asymmetrical, with synclinal limbs that dip gently to the south but plunge steeply to the north. In the Rathcormack syncline, for instance, the strata of the northern limb have a maximum dip of 40°, whereas those on the southern side of the syncline are in many places vertical. Similarly, in the North Cork syncline the strata of the northern limb at Dunmahon have dips of

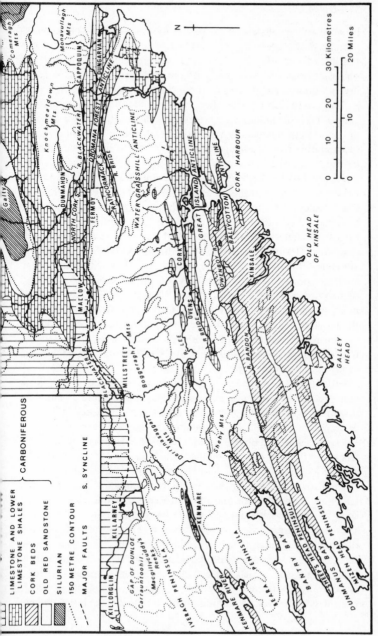

Fig. 3.5 The Ridge and Valley Province.

CARBONIFEROUS

LIMESTONE AND LOWER
LIMESTONE SHALES

CORK BEDS

OLD RED SANDSTONE

SILURIAN

150 METRE CONTOUR

MAJOR FAULTS

S. SYNCLINE

between 10° and 40°, whereas on the southern side of the syncline, near Fermoy, the rocks are sharply overfolded and the strata dip southwards at 70° (Monkhouse 1964). In chapter 5 it will be suggested that this asymmetry of the structures could have some bearing upon the character of the region's drainage pattern.

But asymmetrical or not, there can be no doubt but that the folds control both the pattern of the region's geological outcrops and the distribution of its uplands and lowlands. Basically the region can be thought of as a great corrugated sheet of Old Red sandstone exposed by the widespread removal of the former Carboniferous mantle. Northward of the Cork Harbour to Kenmare line both the Lower Limestone shales and the limestone survive chiefly on the floor of linear valleys developed on the region's deepest Old Red sandstone synclines. Apart from these large synclinal outliers, the Lower Limestone shales and the limestone still exist only as a few lenses set into small flexures on the limbs of the major Old Red sandstone anticlines. Small though these lenses are, they do afford clear evidence of the former continuity of the Carboniferous cover. Southward of the Cork Harbour to Kenmare line there is rather more tangible evidence of the former Carboniferous cover. There, in the province of the Cork Beds, the Carboniferous rocks seem to have proved more resistant than have their counterparts farther to the north, and as a result the Cork Beds still blanket the Old Red sandstone over wide areas.

Although the region possesses a clear structural unity, it is convenient to divide it into two portions by means of an arcuate line drawn from Galley Head on the south coast of Co. Cork, along the eastern foot of the Shehy Mountains and the southern foot of the Derrynasaggart and Boggaragh mountains to Mallow in northern Co. Cork. To the east of this boundary the landscape is predominantly one of alternating Old Red sandstone ridges and Carboniferous-floored lowlands, but to the west of the boundary Carboniferous rocks are largely absent and the Old Red sandstone rises to form some of Ireland's highest and most rugged mountain ranges.

In the eastern portion of the region, and northward of the area occupied by the Cork Beds, the ridge-and-valley character of the topography is very pronounced because there the strata are tightly folded, and the closely spaced anticlines and synclines yield a structurally controlled topography of short wave-length. Long, narrow synclinal valleys, containing on their floors the last remnants of the region's former Carboniferous cover, alternate with east-to-west trending anticlinal uplands developed in the Old Red sandstone.

The synclinal valleys are readily identified upon virtually any map of Ireland; their floors rarely rise above a height of 60 m, and they are drained by easterly flowing strike streams such as the Blackwater, the Lee and the two Bride rivers. In those synclines where the Lower Limestone shales and the limestone are both present, it is usually the shales which form the lowest ground, and the strike streams therefore tend to flow not along the limestone but along the narrow shale outcrop immediately at the foot of the bounding Old Red sandstone slopes. This behaviour is exemplified by the Blackwater above Cappoquin and by the Bride in the Rathcormack syncline. But much more remarkable is the manner in which the subsequent rivers of the region suddenly abandon their strike valleys in order to gain the sea by traversing a number of anticlinal sandstone uplands. This drainage phenomenon has attracted the puzzled attention of geomorphologists for over a century; it is a topic to which we will return in chapter 5. In most places on the floor of the synclines the limestone — like the shales — is hidden beneath a drift mantle but there are, nevertheless, a number of minor sinks and risings together with some small caves such as those at Ovens and Castlepook in Co. Cork (Coleman 1965). Along the margins of the strike valleys there is almost everywhere an abrupt break of slope marking the geological boundary where the tough Old Red sandstone rises steeply away from the Carboniferous strata in the synclines. Here we have juxtaposed rocks offering very different degrees of resistance to denudation, and it seems possible that the slopes developed upon the Old Red sandstone in Waterford and eastern Cork are in many localities virtually the stripped Old Red sandstone/Carboniferous interface.

Turning now to the anticlinal Old Red sandstone ridges, we find there a considerable diversity of form. Some of the ridges — those developed upon the Dromana Forest and the Great Island anticlines, for instance — rise only a few tens of metres above the floors of the adjacent synclines, but elsewhere, even in this eastern portion of the region, the Old Red sandstone rises well over 600 m to form the impressive masses of the Knockmealdown, Comeragh and Monavullagh mountains. But one characteristic almost all the sandstone uplands display in common: on either their crests or their flanks they support a display of planation surfaces that is unparalleled anywhere else in Ireland. From the summit of Dyrick [S075058], in the Knockmealdowns, the observer sees beneath him a bench standing at around 210 m and cut into the sandstone of the Knockmealdown anticline, while to the south the entire skyline is occupied by a procession of accordant summits lying again at around 210 m and bevelling the whole broad crest of the Watergrass-hill anticline. Similarly, take the road south from Cork to the city's airport and what greets the geomorphic eye is another set of remarkably accordant summits, here standing at around 135 m and faceting the sandstone of the Great Island and Ballycotton anti-clines. But further discussion of these upland surfaces must be reserved for chapter 4.

In some of the smaller anticlines the Old Red sandstone has experienced only slight denudation and its exhumation from beneath the Carboniferous cover must have been a comparatively recent event. A case in point is the Dromana Forest anticline; there the crest of the anticline is still developed in the comparatively thin Kiltorcan Beds which form the uppermost member of the Old Red sandstone sequence. In other places, however, the Old Red sandstone anticlines have been deeply denuded, and both in the country eastward of the Comeragh and Monavullagh mountains, and in the Slievenamon ridge, the sandstone has been peeled away completely to expose the much less resistant Lower Palaeozoic rocks lying at the core of the folds. The result is some striking topography produced by differen-tial denudation. The Ordovician and Silurian rocks exposed to the east of the Comeragh — Monavullagh massif form the undulating surface of the Rathgormuck Plateau standing at between 120 and 200 m and overlooked by Old Red sandstone uplands in the east, north, and west. The sandstone outcrops in the east and north are only narrow, so the uplands that developed are merely chains of low hills such as the Coolnamuck Hills south of Carrick-on-Suir. But to the west there lies a major Old Red sandstone outcrop and this forms the craggy eastern face of the Comeragh and Monavullagh moun-tains. There, on the flanks of the Comeraghs, is to be found an impressive group of cirques including among their number the gloomy, moraine-stopped bowl of Coumshingaun which is one of the finest Irish examples of that particular landform type. In the Slievenamon ridge, lying a few kilometres to the north of Carrick-on-Suir, the whole Silurian heart has been eaten out of the inlier to form an upland basin drained by the Lingaun River and almost completely surrounded by a narrow rim of hills developed in the Old Red sandstone. More than a century ago the Geological Survey termed this upland basin the Nine-mile-house tableland, and they aptly likened the entire Slievenamon inlier to 'a huge half-consumed pasty' in which the sandstone forms the pastry while the Silurian strata represent the meat within. Nature's pastry is tough; the great conical sandstone peak of Slievenamon rears its head almost 500 m above the Silurian tableland produced by the Gargantuan bite.

In the far south of Co. Cork there lies the country developed in the Cork Beds, but although the geology may differ from that of the more northerly portion of the county, there is little difference to be discerned in the topography. The Cork Beds give rise to the same terrain of synclinal valleys and anticlinal uplands, and those uplands are bevelled by planation surfaces identical in character to the surfaces that facet the Old Red sandstone structures farther to the north. Indeed, nowhere are the surfaces better developed than around the drowned mouth of the River Bandon at Kinsale.

In Waterford and east Cork the Carboniferous-floored synclines are gradually pinched out by the westward rise of the fold axes, and in the western portion of the Ridge and Valley region the Carboniferous rocks are very restricted in their outcrop. There, to the west of the Galley Head to Mallow line, the landscape is completely dominated by the rugged Old Red sandstone mountains which are one of Ireland's chief scenic attractions. It is, nevertheless, the Armorican structures which still remain the key to an understanding of the topography. The folds here have swung around slightly to strike towards the west-south-west, but again it is the anticlines that form the uplands while the synclines underlie the major lowlands. Indeed, the relationship goes further than this because the amplitude of the folding diminishes towards the south and this is reflected in a general southward reduction in the altitude of the sandstone ridges. Towards the west the mountains persist in their strike trend, and there they push out into the Atlantic to form four long peninsulas — the Mizen Head, Sheep's Head, Beara (or Caha), and Iveragh peninsulas — while slightly farther to the north lies a fifth major peninsula — the Dingle peninsula — which is separated from the main body of the Ridge and Valley region by the limestone lowland around Killorglin. Each of these five peninsulas is essentially an anticlinal ridge, but in detail their structure is complex. The outline of the Sheep's Head peninsula, for instance, is determined by strike faults, while in the Beara peninsula the major anticline supports numerous secondary flexures so that some of the peninsula's headlands are anticlinal in structure while others are synclinal. Conversely, of course, the rias that intervene between the major peninsulas are essentially synclinal in character, and it is around the shores of the rias that there outcrop the only Carboniferous strata to have survived in the southwest. One other 'outcrop' in the southwest deserves a passing mention. Around the head of Bantry Bay there occur the members of Ireland's southernmost drumlin-swarm and among the drumlins, but on the floor of the bay, there lies a series of freshwater silts which extends down to a depth of ⁻57 m (Stillman 1968). The silts have been ascribed a tentative age of 11-12000 years B.P., so clearly the most recent submergence of the rias is by no means a distant event.

In characteristic Irish fashion the region's sandstone uplands are fragmented into a large number of short, narrow ranges, although here the intervals between most of the ranges are marked by cols rather than by the intermontane lowlands which feature in other of Ireland's fragmented uplands. This fragmentation of the topography was doubtless the work of fluvial processes acting upon the corrugated surface of the Old Red sandstone, but there can be no doubt that the region owes the greater part of its topographical detail to the activities of a very different agent — ice. Evidence of intense glaciation is to be found almost everywhere in features ranging from roches moutonnées and ribbon lakes, up to the cirques and arêtes that impart a serrated character to many of the skylines. Considerations of space preclude discussion of the features of individual mountain groups, but mention must be made of Macgillycuddy's Reeks which tower above the waters of glacially scoured Lough Leane at Killarney. A familiar tourist-trail leads across the mountains through the glacial diffluence col of the Gap of Dunloe, but there are very few visitors who climb to the crest of the ridge there to set foot upon an impressive array of peaks that includes Carrauntoohil (1050 m), Ireland's highest peak, together with all three of the only other Irish mountains to surpass a height of 965 m.

One region in the western portion of the Ridge and Valley Province is unique — the Dingle peninsula. There, and nowhere else in the southwest, the anticlinal Old Red sandstone has been denuded sufficiently deeply to expose both Lower Palaeozoic strata and the controversial Dingle Beds, which are perhaps of lower Old Red sandstone age. Generalisation is difficult in an area of such geological complexity (some five secondary anticlines and synclines traverse the peninsula) but here, as in Co. Waterford, it would seem that the Old Red sandstone is the most resistant member of the anticlinal sequence. Almost without exception its outcrops give rise to mountainous topography, and particularly striking is the great slab-sided sandstone mass of the Slieve Mish Mountains which rises so abruptly above the limestone of the Listowel-Killorglin Lowland. It is, nevertheless, the Dingle Beds that form the peninsula's most imposing peak — Brandon Mountain (962 m). Brandon offers a spectacular display of coastal scenery where the slopes of the mountain fall precipitously towards the Atlantic, and it is also notable for an impressive range of remarkably fresh glacial and periglacial features. But then the Dingle peninsula as a whole bears striking testimony to the effectiveness of the Pleistocene processes, and it is very fitting that one of the Dingle cirques — Lough Doon [Q505060] standing by the Conair Pass — should in 1848 have been the first such Irish feature to be recognised as the former seat of a glacier.

The southern mountain inliers (4)

Northward of the Ridge and Valley Province there lies the area where the Armorican earthmovements produced a series of open folds developed chiefly along pre-existing Caledonian lines. These folds bring five large mountain-forming inliers up through the Carboniferous limestone of the southern Central Lowland: the Slieve Aughty, Slieve Bloom, Galty, Slieve Bernagh, and Silvermine inliers, the two latter inliers being linked together by a narrow lowland outcrop of Old Red sandstone around the river Shannon (see fig. 3.2, p.19). Geologically the inliers are all similar. They consist of large southwestward or west-south-westward striking periclines containing an outcrop of Silurian rocks set within a frame of Old Red sandstone and basal Carboniferous sediments.

Again the relative weakness of the Carboniferous rocks is in striking evidence because in most places the topography developed upon the inliers rises steeply from the surrounding Carboniferous lowlands to attain heights of around 450 m. The abrupt break of slope at the mountain-foot lies almost everywhere within two kilometres of the Devonian/Carboniferous junction, although, as was suggested earlier, feature mapping by the Geological Survey may locally have brought the geological boundary and the topographical break of slope into undeservedly close coincidence. Within the inliers there are many high mountains developed upon the Silurian strata, but here again there is a tendency for the Old Red sandstone to form the higher and bolder topography. The relative weakness of the Silurian strata is most strikingly demonstrated in the Galty inlier where the Silurian core has been reduced to an upland-girt basin — the Ballylanders basin — overlooked in the east and west by the extremely steep inward-facing Old Red sandstone slopes of the Galty and Ballyhoura mountains. Topography produced by the same differential denudation, but on a very much smaller scale, is to be seen in the Silvermines inlier where small outliers of the former Old Red sandstone envelope survive to form prominent caps atop Keeper Hill and Devilsbit Mountain. On the latter mountain it is the peculiar notch intervening between two sandstone cakes which in legend was supposed to have been produced by some diabolical agent, thus affording the peak its name.

Although the inliers are basically all of the same structure, the Galty inlier merits a few further words because its structure is complicated by the Glen of Aherlow. The glen is a limestone-floored valley 17 km in length and lying between the tor-crowned Galty Mountains in the south and the narrow, steep-sided Slievenamuck ridge in the north. At its eastern end, where it merges with the Central Lowland, the glen is a synclinal trough, but in its upper reaches it is a fault-angle valley, the fault lying on the southern side of the glen where it brings the weak limestone into contact with the much tougher Silurian rocks at the core of the inlier. Another, much larger fault — the Slievenamuck Fault — runs along the northern side of the Slievenamuck ridge where the Carboniferous strata are downthrown against the Old Red sandstone. The resulting topographical feature — a bold escarpment 12 km long and some 150 m high overlooking the limestone lowland around Tipperary — is a splendid example of a fault-line scarp.

The Castlecomer and Slieveardagh Plateaux (5)

Near the southeastern margin of the Central Lowland there lie two extensive outliers of Upper Carboniferous strata, and there they form two major plateaux: the large Castlecomer Plateau and the smaller Slieveardagh Hills (see fig. 3.2, p.19). In both plateaux the strata rise to horizons of sufficient stratigraphical height for a few coal-seams to be present, and in days gone by the two plateaux were among Ireland's chief coal-producing regions. But by international standards the output was trivial and all production has now ceased.

The Castlecomer Plateau covers more than 500 km² and its rocks constitute a geological basin. The gentle centripetal dip of the strata is closely reflected in the topography, and the somewhat drab, monotonous surface of the plateau falls from a height of some 300 m around the rim, to about 150 m at the centre in the vicinity of Castlecomer itself. The surface of the plateau is broken by a few small outward-facing scarps developed upon some of the more resistant beds, and in the west there are a number of low parallel ridges following the line of some minor north-to-south flexures in the underlying strata. The edge of the plateau is steep almost everywhere, and the escarpment is particularly impressive in the east where it overlooks the Barrow valley between Carlow and Muine Bheag. In this section the ground falls from 330 m on the plateau, to 90 m in the Barrow valley, all within a distance of less than 2 km. At the southwestern margin of the plateau there lies one of Ireland's best-known caves — the Dunmore cave [S509651] — developed in the limestone just below the base of the Upper Carboniferous strata (Dunnington and Coleman 1950; Coleman 1965).

The very much smaller plateau of the Slieveardagh Hills is separated from the Castlecomer Plateau by the lowlands of the Nore valley. The geological structure of this second plateau is generally similar to that of its northern neighbour, but the Slieveardagh basin is more tightly folded along a northeast to southwest axis, and the basin is asymmetrical, the dips on its northwestern side being much steeper than those along the southeastern margin. As a result of this asymmetry, the plateau assumes the topographical form of a large cuesta. The southeastern slopes of the plateau are gentle and broken only by a few low, discontinuous scarps developed upon some of the more resistant strata; the plateau-surface proper is a narrow, gently undulating upland standing at around 270 m; and the northwestern slopes constitute an escarpment more than 120 m high overlooking the Central Lowland around Urlingford.

The Abbeyfeale Plateau (6)

To the southwest of the Central Lowland, and extending from the

Ridge and Valley Province northwards to the shores of the Shannon estuary, lies Ireland's most extensive outlier of Upper Carboniferous strata. The rocks present are dark shales, siltstones and sandstones (despite the Geological Survey's mapping of them as Coal Measures, there are very few coal-seams present) and they form a large plateau structurally comparable with the Castlecomer and Slieveardagh Plateaux. The moorland surface of the plateau, with its heavy gleyed soils and thick banks of peat, varies in height from 90 to 340 m, and above this surface there rise a few subdued hills bearing such rather grandiose titles as the Mullaghareirk Mountains and the Stack's Mountains. Northward of Abbeyfeale the plateau is dissected by the deep valleys of such westward-flowing rivers as the Galey and the Allaghaun, both of them following the strike of some gentle east to west folds, but southward of Abbeyfeale the heart of the plateau is an area of centripetal drainage focused upon the River Feale. Along much of its margin the plateau is edged with a series of narrow terraces developed upon some of the more resistant members of the Upper Carboniferous sequence, but in the east, from Dromcolliher almost to the Shannon — a distance of 45 km — the plateau ends in a bold escarpment which locally is more than 220 m high. A similar escarpment occurs in the west around Tralee and Castleisland where the plateau overlooks the Listowel — Killorglin Lowland.

The Listowel – Killorglin Lowland (7)

Lying between Killarney and the Shannon estuary, this sinuous lowland is really an outlying portion of the Central Lowland. Its surface mostly lies within 60 m of sea-level, and it is developed almost entirely in drift-encumbered Carboniferous strata. Where limestone appears at the surface there is a scattering of solution phenomena including a few caves as, for example, around Tralee (Coleman 1965). In only two places is the lowland interrupted by uplands. Firstly, a mass of Upper Carboniferous strata rises to 272 m in Knockanore Mountain immediately to the south of the Shannon, and secondly, in the Kerry Head Peninsula the Old Red sandstone rises to 220 m in forming what, topographically at least, is a miniature version of the Dingle peninsula.

Fig. 3.6   The northern portion of the Clare Plateau (after Sweeting 1955).

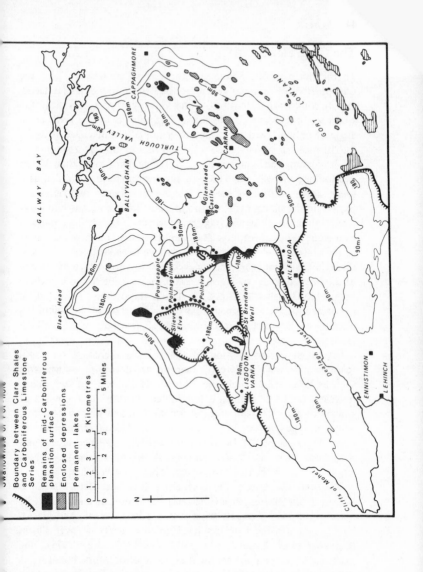

Legend

Swallowhole of Pot-hole

Boundary between Clare Shales and Carboniferous Limestone Series

Remains of mid-Carboniferous planation surface

Enclosed depressions

Permanent lakes

0 1 2 3 4 5 Kilometres
0 1 2 3 4 5 Miles

N

GALWAY BAY

Black Head

CAPPAGHMORE

TURLOUGH VALLEY 90m

90m

180m

90m

90m

BALLYVAGHAN

180m

90m

CARRAN

Glensleade Castle

GORT LOWLAND

90m

180m

Poulacapple

Pollnagollum

Polleva

Slieve Elva

St Brendan's Well

180m

90m

KILFENORA

180m

90m

180m

LISDOON VARNA

90m

Dealagh River

ENNISTIMON

LEHINCH

180m

90m

Cliffs of Moher

Clare Plateau (8)

.. the north of the Shannon estuary lies Co. Clare. There the Carboniferous strata may rest upon the firm foundation of a deeply buried southward extension of the Galway granite; certainly the strata are little disturbed. Regionally they dip southwestwards at an angle of only a few degrees, and they underlie a dissected plateau, the surface of which reflects the geological dip by rising from sea-level along the Shannon estuary, to heights of more than 300 m overlooking Galway Bay. Indeed, the region can be regarded as a great, northward-facing cuesta. Rocks belonging to two series are present in the plateau: the Carboniferous limestone, and the overlying Namurian shales and flagstones, the two series being separated by a plane of disconformity. The Namurian rocks outcrop over most of the plateau, but in the north they have been stripped away and there the limestone is extensively exposed in an area which offers one of the finest displays of limestone landforms to be found anywhere within the British Isles (fig. 3.6). The juxtaposition of this karstic region against the ill-drained, peat-covered surface of the Namurian portion of the plateau affords one of Ireland's most striking landscape contrasts.

In the Namurian division of the plateau there are few bold features and even the plateau's margins are devoid of escarpments such as those which limit the Abbeyfeale Plateau. In one place, however, the Clare Plateau is terminated abruptly: to the north of Liscannor Bay the Namurian strata have fallen prey to Atlantic waves and the Cliffs of Moher fall sheer to the sea from a height of up to 196 m (see p.193). Inland from the cliffs, and especially around Kilfenora, there are large numbers of drumlins, most of them aligned from northeast to southwest (Finch and Walsh 1973).

The limestone area at the northern end of the Clare Plateau is known as the Burren, from the Irish 'boireann' meaning 'a rocky place'. And that is exactly what it is — a bare, karstic and glacio-karstic region rising from sea-level up to a height of over 300 m (Sweeting 1955; Tratman 1969; Williams 1970; Drew 1975). The existence of such an upstanding mass of solution-prone limestone is an anomaly in Ireland, and it seems that the Burren limestone survives at its present high level only because it retained its protective roof of Namurian strata until as recently as glacial times. One large outlier of the Clare shales — the lowest member of the Namurian series — still lies on the limestone around Poulacapple [M185044] to the northeast of Lisdoonvarna, and in this outlier, as elsewhere, the Clare shales (they are about 38 m thick) form a low escarpment, their black colour contrasting sharply with the grey-white limestone beneath. In other places the removal of the Namurian beds was an event so recent that the modern landscape still contains fragments of the exhumed intra-Carboniferous erosion-surface upon which the shales once rested. Such fragments exist around the Poulacapple outlier, to the north of Slieve Elva, and in many another locality (see fig. 3.6). Since the unroofing of the limestone was a very recent event, it follows that the karstification must also be modern, and this view is supported by the fact that the Burren's system of ground-water circulation is still very poorly organised. Similarly, many features indicate a youthful age for the region's caves. But has the Burren been an area of bare limestone throughout post-glacial time? It is too early to answer that question with confidence, but it does seem probable that the limestone was formerly mantled with drift, and that overgrazing later resulted in vigorous soil erosion which laid bare the underlying strata.

The eastern, northern, and western limits of the Burren are marked by steep escarpments which in many places have been smoothed and rounded by glacial abrasion. But the most striking feature of these escarpments is their terraced character as tier upon tier of scars and limestone pavements reflect the gentle dip of the bedding-planes. Williams (1966) regards the pavements as glacio-karstic features formed where post-glacial solution has been at work upon the sheets of glacially scoured limestone that were left behind as the ice retreated. On and around the pavements innumerable streams disappear along the junction between the Namurian strata and the limestone, and a particularly fine series of swallets is located along the eastern slopes of Slieve Elva. Dry valleys abound, the two largest being the Ballyvaghan and Turlough valleys which deeply trench the plateau on its northern side. Another fascinating dry valley lies immediately to the east of Lisdoonvarna. There, above St Brendan's Well [R145985], a dry valley is developed virtually along the plane of the intra-Carboniferous erosion-surface, the floor of the valley being in the limestone while its walls are developed in the Clare shales. Most of the region's subterranean drainage follows the southward dip of the limestone via small caves and fissures, but the region does contain one system of large caves in the Pollnagollum —

Pollelva complex lying to the east of Slieve Elva (Coleman and Dunnington 1944; Coleman 1965; Tratman 1969). The galleries of that system have now been explored for a distance of more than 11 km, and this makes it the longest cave-system known in Ireland. Another of the Burren's well-developed features is the enclosed depression, some of the depressions being due to cavern collapse, while others are the result of surface solution. On low ground, and especially along the shores of Galway Bay, a few of the depressions are turloughs subject to periodic flooding. Some good examples of depressions are to be seen from the main road in the vicinity of Glensleade Castle [M228010]. Most hollows vary in size from only a few tens of metres in diameter, up to the large Carran depression [R280990] which is over 3 km long and more than 1 km wide. That particular depression is elongated from northeast to southwest along the axis of a shallow anticline, and its floor — partly pavement, partly moraine, and partly turlough — lies some 70 m below the level of the surrounding plateau. The feature has been claimed as the best example of a polje within the British Isles (Sweeting 1953). It is probably a pre-glacial feature formed by the coalescence of a group of smaller depressions which, in their turn, had been shaped by acid waters draining off some local Namurian outlier — an outlier which was perhaps finally obliterated as a result of glacial erosion.

The three Aran Islands — Inishmore, Inishmaan and Inisheer — lying off the Clare coast are composed of the same gently dipping limestones as form the Burren, and the islands may be regarded as outliers of the northern Clare Plateau. The islands are almost completely devoid of surface-water — even the storage of rain-water presents problems — and although their array of features such as swallow-holes and enclosed depressions can hardly match that of the mainland, the three islands do afford a magnificent display of a wide variety of different types of limestone pavement (Langridge 1971). The terraced effect imparted to the landscape by the pavements is particularly well seen in Inishmore. There the topography rises gradually from the northern coast in a series of treads and risers before ending abruptly along the southern coast in a range of overhanging cliffs.

The Iar-Connacht Lowland (9)

To the north of Galway Bay there lies an extensive lowland

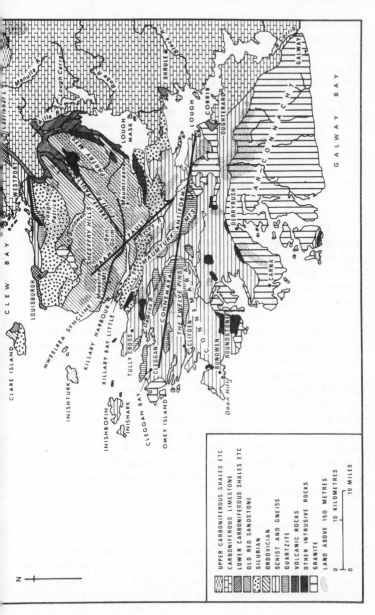

Fig. 3.7 The Iar-Connacht Lowland and the Killary Mountains.

developed chiefly upon the Caledonian Galway granite, but in places extending northwards onto the pre-Cambrian rocks of Connemara and onto such other ancient, but lesser granite bodies as those of Roundstone and Oughterard (fig. 3.7). The lowland mostly lies below a height of 120 m and at least one geologist (Burke 1957) has viewed it as an exhumed sub-Carboniferous erosion-surface. Whatever its origin may be, there can be no doubt but that the lowland today owes its character to glaciation. It is an area of knock-and-lochan topography — a wilderness of bare moutonnée surfaces and huge frost-shattered boulders; of ice-scoured rock-basins and heather-clad moraines; of kilometre upon kilometre of sodden blanket-bog littered with innumerable sheets of brown, peat-stained water. Above the lowland there rise some steep-sided residual hills and mountains, and of these the most interesting is Doon Hill [L593425], near Bunowen, developed upon a volcanic neck of presumed Tertiary age. A relationship between the residual features and the underlying geology may exist in many another locality because the Galway granitic complex is far from being homogeneous. In the west, for example, in the Carna region [L785320], five distinct granite types have been identified within an area of only 36 km² (Wright 1964), but elucidation of the region's structural geomorphology must await further study. In many places the granite certainly seems very susceptible to weathering. The post-glacial release of ice-pressure has perhaps been responsible for transforming many a moutonnée surface into a chaos of broken blocks, while elsewhere granite core-stones are to be found lying *in situ* and set within a matrix of rotten granite or growan. A good place to inspect such a section is in a quarry at Derryrush [L884390], while the mountain to the southwest of the quarry affords a fine view over the surrounding lowland. On the way to the summit there are to be seen possible altiplanation terraces, glacial crescentic gouges, and the scars left in the blanket-bog where saturated peat has become mobile and moved away downslope.

The Killary Mountains (10)

On the borders of counties Galway and Mayo, and to the west of loughs Corrib and Mask, there lies a rugged, mountainous area centred upon the long fiord-like inlet of Killary Harbour. The remarkable fragmentation of the uplands in the immediate vicinity of Killary Harbour was mentioned in chapter 2 (p.12), and in this region as a whole the mountains form fifteen compact units which rise steeply above the intervening lowlands (fig. 3.7). Seven of the units attain heights of between 300 and 600 m, but the remaining eight units all rise above the 600-m contour, the highest peak in the region being Mweelrea at 827 m. Among the units are such uplands as the Sheeffry Hills, the Mweelrea Mountains, the Maumtrasna massif, the Partry Mountains, the Maumturk Mountains, and, best-known of them all, The Twelve Pins of Connemara.

The region possesses a great geological complexity, but it consists essentially of a pre-Cambrian and Cambrian basement overlain by thick sequences of Lower Palaeozoic strata. The basement rocks are the Irish Dalradians — a highly metamorphosed mass consisting in Co. Galway chiefly of the Connemara schists, and penetrated here and there by such intrusive bodies as the Omey and Oughterard granites. Apart from true mica-schists, the basement includes a variety of rock-types, including the ophicalcites which are so familiar to the souvenir-hunting tourist as Connemara marble, but topographically the most significant rocks among the schists are some thick and extensive beds of quartzite. The main outcrop of the basement is in the Connemara antiform lying in the southern portion of the present region, but the schists reappear to the north in a very narrow belt of country close to the southern shores of Clew Bay. The actual shores of the bay, however, are underlain by younger Palaeozoic strata, including Old Red sandstone and Carboniferous limestone, these rocks having been downthrown against the schists by the Clew Bay Fault. In the county between the two schist outcrops, the basement is buried beneath a great pile of Lower Palaeozoic strata lying chiefly within the Mweelrea-Partry syncline. These rocks are Ordovician and Silurian in age, and they extend throughout the greater part of the barony of Murrisk which lies between Killary Harbour and Clew Bay. Complex and diverse though the geology of the Killary Mountains region may be, there is one feature that all its rocks display in common, whether they belong to the basement or to the Lower Palaeozoic cover: they all reveal the clear impress of the Caledonian orogeny, the fold-structures striking southwestwards in the northeast of the region and from there swinging westwards.

In the area of the Connemara schists there is a close relationship between geology and topography. The quartzites form mountain units such as the Maumturk Mountains and The Twelve Pins, while the remaining members of the sequence underlie the surrounding lowlands. In the area underlain by the Lower Palaeozoic strata the structural control of the topography is perhaps less dramatic, but it nevertheless remains important. In this context four points deserve to be made.

(1) The entire grain of the country reflects the underlying geological strike as it swings from southwesterly in the Partry Mountains, to westerly in the western Sheeffry Hills.

(2) There is evidence of differential denudation among the Lower Palaeozoic rocks. In the Ordovician succession the sandstones and ignimbrites of the Mweelrea group seem to be particularly tough and they form the Mweelrea and Partry Mountains, while on the other hand, the east to west trending Glenummera valley, lying between the Sheeffry Hills and Ben Gorm, is associated with the presence of the intensely cleaved slates of the Ordovician Glenummera formation. In the northwest of the region Silurian strata form the lowland south and east of Louisburgh, while a few kilometres to the east of Louisburgh there is a superb example of the structural control of topography where Silurian quartzites form the great conical peak of Croagh Patrick rising 773 m above the drumlin-filled waters of Clew Bay.

(3) Some of the many faults present in the region have considerable topographical influence. The Erriff valley at the head of Killary Harbour, for instance, is developed along a major fault — the Erriff Fault — and the depression which can be traced south and southeastwards from Doo Lough, across Killary Harbour, and thence along the Joyce's River to Lough Corrib is similarly controlled by the presence of the Maam Faults. Killary Harbour itself is also largely a fault-line feature overdeepened by ice-action, although the development of this remarkable trench owes something to the presence of an outcrop of the relatively weak Glenummera formation. On a lesser scale the influence of faulting is reflected in a number of small fault-line scarps such as that developed along the Salrock Fault at the eastern end of Killary Bay Little [L780640]. The slab-sided seaward face of the Croagh Patrick range may also be a fault-line scarp associated in this case with the Clew Bay Fault which lies a kilometre or so away to the north.

(4) Dewey and McKerrow (1963) suggest that the remarkably

smooth plateau surface cut across the Ordovician rocks of the Maumtrasna massif at a height of some 680 m is essentially the exhumed and slightly domed sub-Carboniferous surface. In support of this interpretation they draw attention to the presence of two very small outliers of basal Carboniferous sandstone upon the surface, and their interpretation is one that will be returned to in chapter 4 (p.87).

Some of the region's highest peaks — Croagh Patrick, for example, or the Mweelrea Mountains — were perhaps nunataks throughout even the most intense of the Pleistocene's glacial phases, and Synge (1969) has suggested that the greater part of Murrisk was ice-free during the last glaciation. Glacial relics are, nevertheless, widespread throughout the region (Orme 1967; Synge 1968). Cirques, U-shaped troughs, and moutonnée forms are all well represented and the almost ubiquitous cover of till is revealed in innumerable roadside cuttings. Even drumlins are present as, for example, near Tully Cross [L700617] and on the southern shores of Cleggan Bay [L600585]. The post-glacial interval has left its legacy in the form of extensive peat deposits (they are particularly extensive in the Silurian country south of Louisburgh) and the soil map of Ireland represents the entire Killary Mountains area as one of peaty gleys.

Northwestern Mayo (11)

To the north of Clew Bay, and largely contained within the barony of Erris, there lies a remote area developed in pre-Cambrian schists and gneisses of Lewisian, Moinian and Dalradian ages (fig. 3.8). But although rocks of many varieties are present, it is again the tough Dalradian quartzites which dominate the landscape to form a series of prominent mountains rising impressively above the adjacent lowlands. Those lowlands have been bevelled across some of Ireland's most gnarled and ancient rocks, but rarely do those rocks appear at the surface because the lowlands are dank wastes of thick blanket-bog.

In plan the uplands may be thought of as forming a recumbent letter H. One side of the H is the arcuate upland belt of quartzites extending from Lough Conn in the east, to Achill Island in the west. The other side of the H is the chain of hills that follows the northern Mayo coast between Rathlackan in the east and Benwee Head in the west, the western half of the ridge being developed in the Dalradian schist and quartzites, while the eastern half, eastward of Belderg, is fashioned out of the Carboniferous strata of the Ballina Syncline. The cross-member of the H is the gently curving ridge running from Maumakeogh in the north to Glennamong in the south, the northern half of the ridge being formed in the schists and the Carboniferous strata while the southern half, southward of Carrafull and the Owenmore River, is developed in the quartzites which there form the towering peaks of the Nephin Beg range. The fragmentation of all these uplands is almost as marked as the fragmentation within the Killary Mountains region. Many of the upland units of northwestern Mayo contain only one peak, such as Nephin near Lough Conn, or Knockmore and Slievemore in Achill Island, all of them being isolated conical peaks that rise abruptly out of the surrounding lowlands.

Faulting is here perhaps of less geomorphic significance than in the country lying to the south of Clew Bay, but even so many a feature in northwestern Mayo is related to fault-structures. Blacksod Bay follows the path of the major Blacksod Fault, while the bold eastern face of the Nephin Beg range between Corslieve and Maumykelly — a distance of 5 km — is a fault-line feature related to the downthrow of the Carboniferous strata against the quartzites. On a very much smaller scale, many of the sea-caves and geos on the north Mayo coast are either fault-guided or developed along the outcrop of dykes of presumed Tertiary age. The effect of intrusions is again seen to the south of Belderg, where sheets of dolerite intruded into the schists form prominent scarps on the slope of Benmore. Yet another structural feature is encountered in the south, in the Corraun peninsula, where the removal of the Old Red sandstone has revealed a gently dipping, sub-Devonian surface cut across the quartzites. This surface forms a striking local landscape feature to the east of Corraun Hill and on the long spur [L792975] lying to the southwest of Glennanean Bridgs (Flatrès 1954, 1957).

The region abounds in glacial features, but it is the cirques which deserve especial mention. The five cirques on Croaghaun in Achill Island are especially interesting (Farrington 1953b), and one of them — the cirque containing Lough Bunnafreeva West [F570074] — lies at a height of 360 m perched immediately above the magnificent sea-cliff which terminates the mountain on its northwestern side. Continued coastal recession there is slowly undermining the moraine

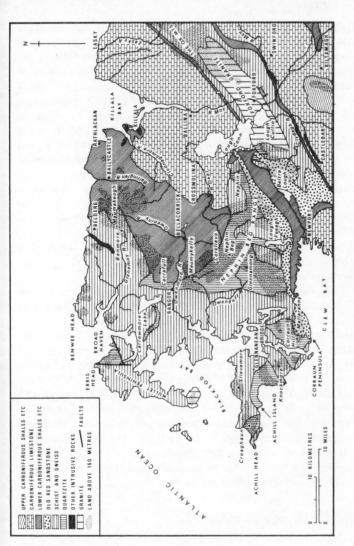

Fig. 3.8 Northwestern Mayo.

that impounds the tarn and resulting in a slow fall in the level of the lake's waters. Farther to the east a fine group of cirques frets the slopes of the Nephin Beg range, and to look northwards from Clew Bay on a summer's evening is to see in the range a good example of biscuit-board topography picked out by the low angle of the setting sun. One other possible glacial feature merits special mention: the deeply incised valley between Bellacorick and Bangor by which the Owenmore River traverses the highlands lying to the north of the Nephin Beg range. In the past this water-gap was usually explained as being a result of drainage superimposition, but modern opinion seems to favour an explanation involving glacial modification. Perhaps the valley assumed its present form under the influence of westward draining glacial meltwaters, and it may be that the Owenmore is a post-glacial creation resulting from drainage on the eastern side of the upland being diverted to the west by morainic deposits in the vicinity of Bellacorick (Synge 1963b, 1968). The anomalous course of the Owenmore certainly presents the geomorphologist with an intriguing problem.

Drift deposits of various types are widespread in the region. Periglacial material is well represented because most of the region escaped the last Irish ice-sheet (see fig. 6.2, p.121). the limits of which are marked by a line of end-moraines extending from Ballycastle in the north, along the eastern slopes of the northern part of the Nephin Beg range, and then southwestwards across the mountains to Mulrany [L825971] on the northern shores of Clew Bay. Dunes of calcareous sand occur at many coastal locations and they are particularly widespread in the southern half of the Mullet Peninsula. But of all the drift deposits present in northwestern Mayo it is peat which will linger longest in the memory of a visitor — a recollection of kilometre after kilometre of sodden blanket-bog, studded with the boles of ancient trees, and spreading across the landscape like some malignant fungus.

The Ox Mountains inlier (12)

Between Newport in the southwest and Manorhamilton in the northeast, there lies the narrow inlier of the Caledonian Ox Mountains anticline. Its length is about 100 km, and its Dalradian and Moinian rocks, plus the Foxford granite, rise through the Carboniferous strata to form a ridge of hills and mountains intervening between the Central Lowland to the southeast and the Donegal-Ballina Lowland to the northwest. In many places the inlier's margins are faulted, and its southeastern limits lie close to what is usually regarded as the southwestward prolongation of Scotland's Highland Boundary Fault. Near Foxford the uplands developed upon the inlier are interrupted by the lowland around loughs Conn and Cullin, and in the north, where the inlier is narrowest, it forms only a scattering of disjointed and glacially scoured hills lying between Collooney and Manorhamilton, the breaks in the range here probably being associated with north to south faults. But elsewhere the inlier forms steep-sided mountain and plateau topography which rises to over 520 m in the Slieve Gamph or Ox Mountains. Especially striking are the bare quartzite hills lying between Coolaney and Ballysadare in north central Sligo, many of the hills being cut by deep, structurally guided north to south ravines known locally as 'alts'. One other feature of the hills to the south of Ballysadare is that they mostly display a type of crag-and-tail form with bold, rocky faces to the north, and drift plasters on the south. But of the region as a whole little more can be said; it remains something of a geomorphic *terra incognita*.

The Donegal – Ballina Lowland (13)

From Dunkineely in the north, to Killala in the south, the shores of Donegal, Sligo and Killala bays are fringed by a lowland lying mostly within 70 m of sea-level. The feature is developed in the Carboniferous strata of the Donegal, Sligo and Ballina synclines, although over most of its extent the lowland carries a thick drift mantle including, at the coast, many extensive areas of sand-dunes. In the north, one branch of the northern Irish drumlin-swarm reaches the sea along the shores of Donegal Bay (Wright 1912), and its presence is clearly indicated by the contour pattern on Half-Inch Sheet 3. In the southern portion of the drumlin region, in the area known as The Pullans, there is to be seen an unusual karstic feature. In the Brownhall demesne [G940700], near Ballintra, the River Blackwater goes underground for several hundred metres, but through a series of pits the river can be seen flowing some 6 m below the surface. Another interesting hydrological phenomenon lies 5 km east of Sligo where Colgagh Lough has no surface link with either Lough Gill or the sea (fig. 3.9). The surface of the lowland is diversified by a few hills formed in reef-limestones, as at Dooney Hill [G721323] on the southwestern shores of Lough Gill, but to the north of Sligo the Moinian inlier at Rosses Point has little geomorphic expression. The most striking feature rising above the lowland is unquestionably Knocknarea, which attains 328 m west-south-westward of Sligo, the mountain being developed upon a Carboniferous outlier. The gently dipping strata of the outlier give rise to structural terraces on the northern and eastern sides of the mountain, and to steep escarpments on the western side, while the topography of the southern slopes is noteworthy chiefly for the presence of the vertical-sided cut known as The Glen [G630333], which is presumably a glacial spillway. The relationship of The Glen to the morainic complex lying to the east around Cloverhill House remains uncertain.

The Cuilcagh Plateau country (14)

Co. Leitrim and its environs contain some of Ireland's most interesting terrain (see fig. 3.9). It is a region of high but compact plateaux, possessed of either precipitous or strikingly terraced escarpments, and separated from each other by deep glens, the lake and drumlin-strewn floors of which lie at least 300 m below the level of the adjacent plateaux surfaces. Again, structure is the key to the morphology. The region's strata are horizontal or very gently dipping Carboniferous sediments, and it is great slabs of these rocks which form the tablelands. There could hardly be finer exemplars of the concept of mountains of circumdenudation. Many of the plateaux have their own individual names — the Benbulbin range, for example, and the Dartry, Bricklieve and Slieve Anierin mountains — and the highest of the plateau-blocks forms the Cuilcagh Mountains (670 m), from which is taken the title here adopted for the entire region. Farther to the east, across the remarkable interlacement of drumlins and water which constitutes the Lough Erne Lowland (a portion of the Central Lowland), there lies the Slieve Beagh upland, another plateau of the same type as those of Co. Leitrim, and that outlier is, therefore, included within the present region.

The effects of differential denudation among the various limestone beds is perhaps best seen in the Bricklieve Mountains, and the slopes bounding that particular plateau display striking terraces

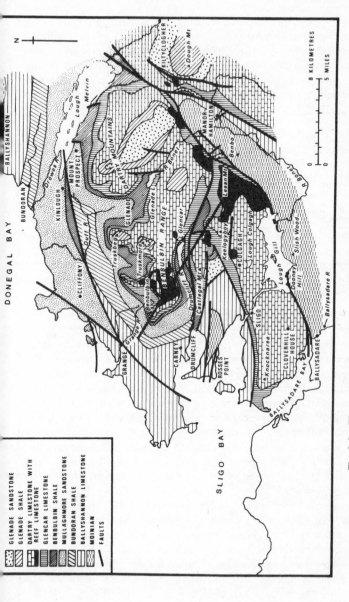

Fig. 3.9 The northwestern portion of the Cuilcagh Plateau country (after Oswald 1955).

Legend:

- GLENADE SANDSTONE
- GLENADE SHALE
- DARTRY LIMESTONE WITH REEF LIMESTONE
- GLENCAR LIMESTONE
- BENBULBIN SHALE
- MULLAGHMORE SANDSTONE
- BUNDORAN SHALE
- BALLYSHANNON LIMESTONE
- MOINIAN
- FAULTS

broken by southeastward trending gullies developed along minor fault-planes. In most other localities the limits of the outliers are marked by impressive mural escarpments which are particularly well developed around the Dartry Mountains and the Benbulbin range. In both these cases the plateaux end abruptly at a height of over 520 m and fine escarpments overlook the coastal lowlands to the west.

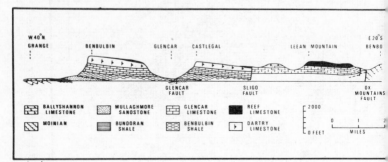

Fig. 3.10 Cross-section of part of the Cuilcagh Plateau country (after Oswald 1955).

In the Dartry Mountains and the Benulbin range, Oswald (1955) has demonstrated the existence of an intimate relationship between geology and topography (figs. 3.9 and 3.10). The shaly Ballyshannon limestone and the Bundoran shale — the lowest members of the region's 1200-m Carboniferous succession — form the coastal lowlands of Region 13 to the west, while the overlying Mullaghmore sandstone gives rise to a marked break of slope which can be traced intermittently from near Kinlough [G816555], southwestwards to the foot of Benbulbin. There it forms a prominent step, rising above the bog-covered lowland developed to the west upon the Bundoran shale. The next highest formation — the Benbulbin shale — crops out at the foot of the major crenellated escarpment which extends southwestwards from Lough Melvin, but the formation gives rise to slopes which rarely exceed an angle of 60°. Next highest in the sedimentary pile comes the Glencar limestone which forms very steep, but grassed slopes. The limestone contains many shale-partings and these give rise to springs lying at the head of deeply excavated gullies. The Glencar limestone is in turn succeeded by the massive and well-jointed Dartry limestone which forms the bare, sheer cliff-faces lying immediately below the plateau-surfaces — cliff faces that are superbly displayed around Glencar and in the classic postcard view of Benbulbin from W.B. Yeats' grave at Drumcliff. The junction of the impermeable Benbulbin shale with the overlying limestones gives rise to a marked plane of lubrication and many landslip features are to be seen, for example, at Mount Prospect [G850540], on the northern face of Tievebaun Mountain [G763509], and in the Swiss Valley [G743441] or the northern side of Glencar. The bog-covered surfaces of the plateaux themselves are diversified by a few mounds and knobs developed upon reefs in the Dartry limestone, and by small escarpments located upon outliers of the still higher Glenade shale and the Glenade sandstone.

This region contains one of Ireland's finest displays of karstic phenomena (Coleman 1965; Williams 1970). The Bricklieve Mountains contain many sink-holes, caves, dry valleys, and pavements; around Colgagh [G750375] is a terrain reminiscent of the cockpit karst of tropical lands; in the Dartry Mountains streams flowing off the Glenade beds disappear down numerous sinks; Polliska near Geevagh [G840165] is one of the deepest pots known in Ireland; and in both the Dartry Mountains and the Benbulbin range pavements and other solution features occur wherever the limestone rises through the peat-mantle of the plateaux-surfaces. The Shannon itself rises at Shannon Pot [H053318] on the northwestern slopes of Cuilcagh Mountain, and only a few kilometres farther to the northeast, across the border into Co. Fermanagh, there lies an important series of caves formed where rivers disappear into the limestone after leaving the plateau's impervious cap-rocks. The best-known of these caves is the Marble Arch cave-system containing a large lake which feeds the Cladagh River, but despite its interest, the limestone morphology of the region as a whole has received remarkably little attention.

The Donegal Highlands (15)

The northwestern corner of Ireland is a land of misty, cirque-sapped mountains, of deep troughs holding secluded ribbon lakes, and of lowlands where glistening roches moutonnées rear above a blanket-bog which threatens their engulfment, the entire region being fringed with a coastline of an intricacy scarcely to be matched elsewhere in Ireland (fig. 3.11). The rocks of the region, like those of western

Connacht, are mostly schists and gneisses of Lewisian, Moinian and Dalradian age, but within this ancient basement there have been emplaced three granite bodies of Caledonian age. Firstly, there is the Main Donegal granite, a complex of granites of slightly differing ages which underlies the coastal margins of northwestern Donegal and strikes thence northeastwards towards Mulroy Bay. Secondly, in the far north, around the shores of Sheep Haven, Mulroy Bay, and Lough Swilly, there lies the outcrop of the small Fanad Head granite. Finally, in the remote country of south-central Donegal, the Dalradian schists are penetrated by the Barnesmore granite. Geologically the entire region is one of great complexity (Pitcher and Berger 1972), and in the last century the Geological Survey was doubtless prudent to defer the mapping of Donegal until the surveying of all the other Irish counties had been completed. But one facet of the region's structure is clear enough: it possesses a pronounced Caledonian grain. The folds and faults responsible for this marked lineation strike southwestwards over the greater part of the region, but in the far southwest, in the Slieve League peninsula, the structures turn towards the west before disappearing into the wastes of the Atlantic.

Topographically the region is dominated by a series of hill and mountain ridges which follow the Caledonian strike from the Inishowen peninsula, southwestwards and westwards to the Atlantic seaboard. Among those ridges the most prominent peaks are developed in Dalradian quartzites which have here proved just as resistant to denudation as have the equivalent quartzites in counties Galway and Mayo. Among the many striking quartzite peaks of Donegal those deserving of particular mention are Slieve Snaght and Raghtin More in Inishowen; Knockalla Mountain in Fanad; Muckish, Aghla Beg, Aghla More and Errigal in northern Donegal; and Slievetooey and Slieve League in the Slieve League peninsula. For two reasons Slieve League must rank as one of Ireland's most interesting mountains. Firstly, it carries near its 598-m high summit two tiny unconformable outliers of what most authorities have regarded as a Carboniferous sandstone, the two affording mute testimony to the prodigious scale of the local denudation (Guilcher 1957). Secondly, half of the mountain has disappeared into the Atlantic; from near the crest those for whom vertigo holds no terrors can gaze down upon the wheeling gulls and foaming breakers far beneath.

The largest of the Donegal uplands — the Derryveagh Mountains —

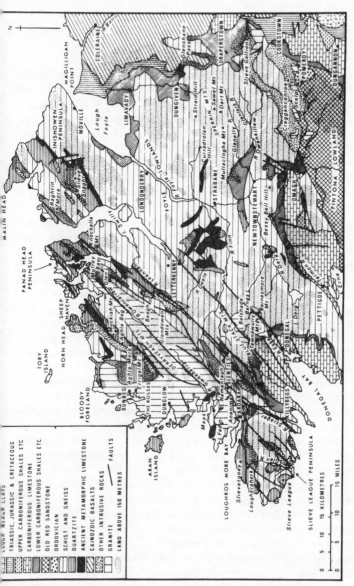

KEY

LOUGH NEAGH CLAYS
TRIASSIC, JURASSIC & CRETACEOUS
UPPER CARBONIFEROUS SHALES ETC
CARBONIFEROUS LIMESTONE
LOWER CARBONIFEROUS SHALES ETC
OLD RED SANDSTONE
ORDOVICIAN
SCHIST AND GNEISS
QUARTZITE
ANCIENT METAMORPHIC LIMESTONE
CAINOZOIC BASALTS
OTHER INTRUSIVE ROCKS
GRANITE
LAND ABOVE 150 METRES
FAULTS

0 5 10 15 KILOMETRES
0   5   10   15 MILES

Fig. 3.11   The Donegal Highlands.

is, nevertheless, developed not in the quartzites, but on the outcrop of the Main Donegal granite, and that fact raises the intriguing but little-understood question of the geomorphic significance of the various Donegal granites. In most cases the granites form a mountainous topography. The Main Donegal granite underlies the Glendowan Mountains as well as the Derryveagh Mountains, and the Barnesmore granite lies at the heart of the Blue Stack Mountains. But elsewhere the Donegal granites form topography of a very different character. The granite of the Rosses ring complex — one of the subdivisions of the Main Donegal granite — underlies the mammilated and boulder-strewn lowland of The Rosses to the north of Dunglow, while the Fanad Head granite forms the coastal lowlands at the tip of the Fanad peninsula and the corridor lying at the foot of Raghtin More in Inishowen. Especially interesting is the case of the Ardara diapir, a circular granite body some 8 km in diameter lying to the north of Ardara and forming the southwestern extremity of the Main Donegal granite (Akaad 1956). The granite has been reduced to a lowland, but the whole grain of the lowland's topography is concentric as a result of denudation having etched out the diapir's foliation, and everywhere, except along the shores of Loughros More Bay, the lowland is overlooked by a circle of hills developed upon the diapir's aureole rocks. The concentric nature of the topography is evident even on Half-Inch Sheet 3, the grain being nicely picked out by loughs Faa, Namanlagh and Ananima.

In fig. 3.11 there are represented some of the many faults which cross the region and profoundly influence its topography, but it should be noted that most of the linear features on the map are really complex zones of fracturing and shattering. These zones have been picked out by fluvial and glacial erosion; they are commonly followed by subsequent rivers, and a few of them have been gouged out to become long narrow trenches. In the latter category is the remarkable trench developed along the line of the Gweebarra Fault for a distance of 48 km. The feature first becomes apparent near Glen [C120310], whence it strikes southwestwards via Glen Lough, the Owencarrow River, and Lough Beagh. It intervenes between the Derryveagh and Glendowan mountains, is picked out by the line of the Owenwee River, and finally reaches Maas [G770982] along the partially drowned valley of the Gweebarra River. Only 6 km away to the southeast the Errig, Finn and Carbane faults underlie the strike valley that runs southwestwards to Glenties, and farther again to the southeast by a similar distance, the important Leannan Fault — often regarded as a branch of Scotland's Great Glen Fault (Pitcher *et al*. 1964) — underlies a variety of features along an 80-km belt of country extending from Lough Swilly southwestwards to near Killybegs. Another smaller, but interesting feature, is located where the southwestward striking Lough Belshade Fault intersects the Barnesmore granite of the Blue Stack Mountains. The fault is a tear which has offset the northern and southern portions of the granite by 3.5 km, with the result that in the northeast a granite fault-line scarp faces northwestwards across a schist lowland, while in the southwest a granite fault-line scarp faces southeastwards across a schist lowland. The Lough Belshade Fault itself is accompanied by a belt of intense crushing up to 100 m in width which has been picked out by the Owendoo River, and it can hardly be fortuitous that the cirque containing Lough Belshade lies upon the same line of weakness (Leedal and Walker 1954). So extensive is the fault control of topography throughout the region that any complete discussion of the fault-guided features is quite out of the question, but a few more examples will further serve to illustrate the genre. To the northeast of Pettigoe [H110670] the Termon River follows the line of the Pettigoe Fault; the rivers Eske and Lowerymore follow the line of the Barnesmore Fault to the northeast of Donegal town; the lowland which bisects the Tawnawully Mountains [H010870] is developed on the Barnes Lough Fault; in the Slieve League peninsula rivers such as the Glen, the Glenaddvagh, Stragar and Owenwee are all fault-guided subsequents (Dury 1964; Reffay 1966); and in the far northeast the shape of ice-gouged Lough Swilly is controlled by the pattern of the local fault-system (Evans 1973).

Faults are by no means the only type of structure to exert a control upon the topography. In the area around Lougheraherk [G592883] in the Slieve League peninsula, and elsewhere, calcitic and dolomitic limestones within the Dalradian sequence have been eroded to form topographical depressions. In the granite areas many a valley and gully have been developed along master joints, and in the vicinity of the Barnesmore granite both crush-zones and Tertiary dykes have all been weathered out to form deep, narrow clefts. Even a large feature such as the Poisoned Glen [B940170] is essentially a subsequent structure guided by Tertiary dykes (Dury 1959).

Although structure controls so many of the features within the

Donegal Highlands, it was the Pleistocene glaciers which were responsible for shaping most of the region's topographical detail (Charlesworth 1924). Evidence of glacial erosion is to be seen everywhere from the cirque-fretted Derryveagh Mountains to the knock-and-lochan terrain of The Rosses. Glacial breaching of pre-glacial divides has been described in the Errigal-Muckish ridge, in the Blue Stack Mountains area, and in the Slieve League peninsula (Dury 1958, 1959, 1964), while the erosive work of glacial meltwaters is to be seen in spillways lying along the margins of the Foyle lowlands in eastern Donegal (Wilkinson *et al.* 1908). One final category of Quaternary feature must be mentioned. In the northern part of the Inishowen peninsula there is a remarkable flight of rock-platforms and raised-beaches standing at up to 25 m O.D. and marking the level of successive post-glacial strandlines (Stephens and Synge 1965). Nowhere else in Ireland is there to be seen such a striking stairway.

The mid-Ulster Highlands (16)

'Strangely diffuse' would seem to be an apt comment upon this region lying chiefly in counties Londonderry and Tyrone. To the west the firm pattern of ribbed uplands gives Donegal its personality, and to the east the Tertiary basalts of Antrim impart an equally strong character to Ireland's northeastern corner, but the intervening region is one where few clear geomorphic patterns can be discerned. Geologically the region could hardly be more varied; its core is an extensive outcrop of the Dalradian schists, but around the margins in the north, east, and south there are rocks ranging in age from the Palaeocene basalts, through the Triassic sandstones of the Lough Foyle shore, the Carboniferous, and the Old Red sandstone, to the pre-Silurian andesites, dolerites, gabbros, and granites of the Tyrone Igneous complex. Topographically the core of the region is the rather subdued upland of the Sperrin Mountains which contains the only peaks in this region to exceed a height of 600 m. Throughout much of the remainder of the region the fragmentation of the uplands has resulted in a confused interdigitation of hills and lowland — a mixing of terrain-types which is well displayed in the west in the landscapes of the Foyle basin around Strabane and Newtownstewart, and in the east in the area lying between The Six Towns and Cookstown. In both these localities the upland units may consist of a single hill or mountain such as Bessy Bell [H390820] or Slieve Gallion [H798880].

The structural control of topography is here far less pronounced than in Donegal — even the Caledonian graining is much less in evidence — and one reason for this is patently obvious on any geological map: those massive beds of Dalradian quartzite, of such significance in Donegal, are largely absent from the country to the east of the Foyle. Mid-Ulster thus contains nothing to compare with Slieve Snaght or the Muckish — Errigal ridge. But that is not to suggest that the structural control of topography is entirely absent. In the north, Lough Foyle occupies a structural basin which was perhaps first downwarped as early as the Carboniferous and then revived by Tertiary diastrophism. In the centre of the region, the gentle arc of the Sperrin Mountains, convex towards the north, reflects the trend of the underlying schists as they swing from a southeastward strike in the east to a southwestward strike in the west (Hartley 1938). Just to the north of the Sperrins, the ridge running from Curradrolan Hill [C480012] to Straid Hill [C590039] is developed upon one of the few quartzite bands present in the schists, and immediately to the south of the ridge the lowland around Inver Bridge is essentially a subsequent feature associated with the outcrop of another member of the schist succession, the outcrop of the Dungiven limestone. To the south of the main Sperrin watershed, rivers such as the Glenelly, the Glenlark, and the Coneyglen Burn are all subsequent streams following the strike of the schists, and in addition the middle reach of the Glenelly valley seems to pick out the course of a southwestward-striking fault. The southern face of the Sperrins, below the peaks of Mullaghclogha, Dart Mountain, and Sawel Mountain, has been widely regarded as a fault-line scarp associated with that same Glenelly valley fault, but whatever its tectonic form may be, the Glenelly valley certainly separates the tough schistose grits that form the High Sperrins, from the seemingly weaker members of the schist series which form the Low Sperrins lying to the south of the Glenelly valley.

Another, and much more important, tectonic line strikes southwestwards across the southern part of the region from near Draperstown and through Omagh. This fault is usually regarded as a part of Ireland's equivalent to the Highland Boundary Fault of Scotland, and although that fault is not usually of any great geomorphic significance in Ireland, here in mid-Ulster its presence is well reflected in the topography. To the northwest of the fault there lie the Sperrins and other uplands developed in the Dalradian schists, while to the southeast of the fault a lowland of subdued form — the Fintona Lowland — is cut across the Old Red sandstone and the rocks of the Tyrone Igneous Complex. Many of the hills that diversify the Fintona Lowland have a structural basis, the hills of Scalp [H635745] and Cregganconroe [H664745], lying to the north-west of Pomeroy, for example, being associated with an outcrop of olivine-gabbros in the igneous complex. The most prominent feature to rise above the lowland is Slieve Gallion, and there the influence of structure is much less easy to discern because the mountain is formed partly in granite and partly in Carboniferous strata. Indeed atop the northernmost of its two peaks there sits an outlier of basalt resting upon chalk. Whatever else it may indicate, that outlier certainly serves as a reminder that both the chalk and the basalt must once have extended well beyond their present confines.

The landscapes of the region may owe something of their subdued character to the fact that the area was completely overridden by the glaciers of the Pleistocene, and its uplands never served as an independent ice-centre (Charlesworth 1924; Colhoun 1970, 1971a, 1971b, 1972). True cirques are therefore absent from the Sperrins, and instead there exist a number of glacially breached watersheds of which the Glenshane Pass [C780035] is one of the most striking. Meltwater channels are also widespread, especially in the country immediately to the north of the Sperrins, and there, too, many a fine section in glaciofluvial sands and gravels is displayed as a result of post-glacial erosion by rivers.

The basaltic plateau (17)

Belfast, like Edinburgh, has a skyline that is dominated by volcanic rocks. From almost anywhere in the city there can be seen to the west and north the commanding peaks of Black Mountain, Divis and Cave Hill, all of them marking the southeastern limits of a basaltic plateau which extends thence for 80 km to the cliffed coast of northern Antrim (fig. 3.12). As we saw earlier (p.5) the basalts are of Palaeocene age, and in most places they rest upon Cretaceous beds — chiefly the flint-banded Upper Cretaceous chalk — which in turn are widely underlain by thin mudstone beds of Liassic and Rhaetic age. But the outcrops of these Mesozoic strata are confined

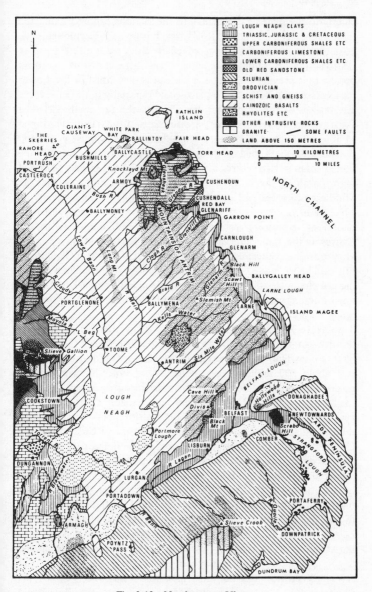

Fig. 3.12 Northeastern Ulster.

chiefly to the plateau's margin — to the sections exposed along the fascinating Antrim Coast Road via Larne and Carnlough, for instance, or in the fine range of chalk cliffs at White Park Bay [D010440] near Ballintoy. Almost everywhere else it is the drab basalts that prevail. Most of the plateau is developed in the Lower basalts (see p.5), but in some places, and especially in the east, there are outliers of the Upper basalts, while in the far north, between Portrush and Ballycastle, there are outcrops of the locally developed Middle (or Tholeiitic) basalts. Those latter basalts must be among the most photographed of all rocks within these islands because the famed columnar structures of northern Antrim are largely confined to the basalts of the middle series. At the Giant's Causeway itself the Middle basalts descend to sea-level as a result of their occupying what seems to be a valley excavated in the Lower basalts during a period of temporary volcanic quiescence, and elsewhere at the Causeway the same interval is represented by the red Interbasaltic Horizon which outcrops so strikingly in the cliffs as it picks out the junction between the Lower and Middle basalts. Spheroidal weathering is a characteristic feature of the horizon and the dark, relatively unweathered basaltic core-stones have for long been known as 'giant's eyes'.

Some 16 km to the east of the Causeway there is an unusual locality — a place where the traveller passes directly off the Tertiary basalts onto Dalradian metasediments, thus spanning with a single step about 500 million years of geological time. Such a journey through time is possible because there, in the northeastern corner of the basaltic plateau, the basalts and the Mesozoic strata have both been removed to expose an inlier — the Antrim inlier — consisting of the Dalradian basement, the Old Red sandstone of Cushendun and Cushendall, and the Carboniferous beds of the Ballycastle coalfield, whence a small output of coal was maintained until very recently. Above the rocks of the inlier there rise a few outliers of chalk and basalt, the most striking of these being the outlier that forms the black-and-white mountain of Knocklayd [D115364] just to the south of Ballycastle. Guilcher (1957) has suggested that elsewhere in the inlier the removal of the overlying Mesozoic and Tertiary rocks may have been an event so recent that some of the topography developed in the Dalradian schists may still in essence be the exhumed sub-chalk surface. But the inlier as a whole has little geomorphic significance; to move off the basalts and onto the Palaeozoic and older rocks is not to experience any very striking landscape contrasts.

The basalts of the northeast have been much disturbed by faulting, but basically the structure of the 'plateau' is that of a trough elongated in a north to south direction — a trough which is clearly of some antiquity because it became a receptacle for the Lough Neagh clays as early as Eocene and Oligocene times. In the plateau today the western lava-flows dip eastwards, and the eastern flows dip westwards, the axis of the trough extending northwards from Lough Neagh towards the coast near Bushmills. This synclinal structure of the flows is clearly reflected in the plateau's topography. In the west the plateau surface climbs gradually to heights of over 350 m before ending abruptly in a bold escarpment overlooking the Roe valley, and in the east the plateau again ascends to over 350 m before being terminated by another and even more impressive escarpment along the fault-guided shores of the North Channel. The plateau might thus be envisaged as consisting of a pair of back-to-back cuestas united along the foot of their respective dip-slopes.

The axis of the structural trough — the junction of the two cuestas — is marked by a lowland standing at a height of less than 60 m O.D., and there, occupying an essentially structural basin, lies Lough Neagh, the most extensive sheet of fresh-water anywhere in the British Isles. Despite its size, the lake is only shallow — most of its waters are less than 15 m deep (Charlesworth 1963c) — and since its surface stands at a height of only 15 m, it is hardly surprising that the land on its immediate margins is often little better than a morass. This is particularly true of the country around the southern half of the lake where the Lough Neagh clays have their outcrop. Late in the Pleistocene, at the time of the Midlandian (Devensian) glaciation, Lough Neagh became considerably swollen when down-wasting ice occupied the Lower Bann valley and blocked the normal outlet of Lough Neagh northwards. Behind this barrier the lake-level rose, flooding the Lower Bann valley until a new temporary outlet was found southwards via Poyntz Pass and Newry, to Carlingford Lough. In due course the melting of the ice allowed the outlet river to establish its present route to the sea along the Lower Bann to Coleraine, but much of the soggy alluvium along the course of the Lower Bann, and around the margins of Lough Neagh, is probably a relic of the lake's swollen late-glacial condition. Even in post-glacial times the lake has been rather larger than it is today, and the greyish diatomite ('Bann Clay' or kieselguhr) found between Toome and Portglenone on the Lower Bann, is a result of sedimentation at that period. One peculiarity of this central region of the plateau is that it is drained by two parallel rivers flowing in opposite directions. As the Lower Bann flows northwards out of Lough Neagh, only 9 km away to the east the River Main is flowing southwards into the lake. Consideration of the region's drainage pattern must await chapter 5, and it suffices here to note that the narrow ridge lying between the two rivers — a ridge often known as Long Mountain — evidently owes its existence partly to faulting within the basalts and partly to their gentle folding.

Around the mouth of the Lower Bann, between Castlerock and Portrush, the central portion of the plateau ends at the coast in an alluvial lowland fringed with sand-dunes, but to the east of Portrush the plateau is in most places terminated by fine ranges of coastal cliffs. Off the north coast, Rathlin Island is merely a detached fragment of the basaltic plateau, and its layered basalt-upon-chalk, black-upon-white, structure is well seen from the sea-front at Ballycastle.

Most of the plateau's surface has a rather subdued character, which is hardly enlivened by a widespread mantle of blanket-bog, although in places there is a marked terraced effect arising from the presence of one or more particularly resistant lava-flows. In some thirty localities, however, the basalts are pierced by olivine-dolerite plugs occupying the site of former volcanic vents, and many of these features form prominent hills rising steeply above the plateau-surface. Slemish [D221053] is the classical example of such a structure, but a second example is Scawt Hill [D335009], and a third is the inclined neck which forms Tievebulliagh [D190266] Two other major Tertiary intrusions of olivine-dolerite merit particular mention, both of them lying in northern Antrim: the Fair Head Sill and the Portrush Sill. The Fair Head Sill is up to 80 m thick and it has been chiefly injected into the Carboniferous rocks of the Ballycastle coalfield. There it forms a capping to the impressive but gloomy promontory of Fair Head, and it gives rise to the great block screes overlooking Rathlin Sound. The much thinner Portrush Sill lies amidst the Liassic shales and there it forms the skeleton of both Ramore Head and the islands of The Skerries. Innumerable lesser intrusions abound throughout the region, many of the dykes standing up to form positive features. This phenomenon is particularly well seen along the coast between Ballycastle and Fair Head because there a number of dykes form causeways running into the sea, the best-known such feature being the much visited North Star Dyke [D143418].

The eastern face of the plateau is spectacularly notched by the deep, U-shaped troughs of the Antrim glens. Some of the glens — there are traditionally nine of them — seem to be associated with lines of faulting, and the two northern glens — Glendun and Glenaan — seem to be picking out the Caledonian grain of the ancient rocks of the Antrim inlier in which both the glens are exclusively developed. But whatever the tectonic basis of the remainder of the glens, they almost all display a common geological structure. The steep upper slopes are formed in the basalts over which streams cascade off the adjacent plateau; in the lower slopes the Cretaceous and Jurassic rocks have their outcrop; and finally, on the floor of the glens, the underlying Trias puts in its appearance. Make a visit to Glenariff, stop south of Kilmore [D225232] and let the eye descend the slopes on the opposite side of the glen. The skyline is developed in the Upper basalts, then comes the Interbasaltic Horizon and the Lower basalts. Next, just above the road encircling the glen, there is the outcrop of the Cretaceous beds, and finally, the Jurassic strata being absent from this site, there are the Triassic rocks underlying the very gently sloping alluvial floor of the glen. There could hardly be a finer illustration of the layered structure of northeastern Ireland.

The glens have clearly been modified by glacial and periglacial erosion, but over the plateau as a whole it is the effects of glacial deposition which most frequently catch the eye (Dwerryhouse 1923; Charlesworth 1939; Stephens *et al.* 1975). Around Ballycastle the Carey and Glenshesk valleys are choked with glacio-fluvial sediments; drumlins are widespread on the lowlands between Lough Neagh and the Armoy moraine; and that moraine, itself a well-marked multi-ridged feature, has at Armoy deflected the River Bush from its original course to Ballycastle and forced it to take a new route to the sea at Bushmills. But much has also been accomplished by post-glacial processes, and in particular the plateau's margins have become classic Irish ground for the study of recent mass-movement. Three types of movement have been differentiated (Prior *et al.* 1968; Prior 1975) and they result in the regular obstruction of traffic on the Antrim Coast Road.

(1) *Slumped blocks* During the Pleistocene, glaciers moving around the margins of the plateau undercut its marginal escarpments. Then, following the withdrawal of the ice, the undercut masses slumped forward, often with a rotation that left a marked slip-scar behind the mobile block. The effects of such movement are superbly displayed on the eastern side of the plateau at Garron Point [D302240] where huge blocks of basalt and chalk have moved over the Liassic clay, and at the northwestern corner of the plateau on Binevenagh [C690300] overlooking Lough Foyle. Interestingly, the slumped blocks nowhere deform the region's post-glacial raised beaches so the blocks must, therefore, have been stable for the greater part of post-glacial time.

(2) *Rockfalls* Small blocks of basalt and chalk regularly fall from the steep slopes that limit the plateau, and the bombardment is such that the traveller on the Antrim Coast Road will soon become familiar with the road-sign advising him to beware of falling rocks. Indeed, at Glenarm [D313155] the road has had to be re-sited on an artificial embankment constructed well clear of the overhanging chalk cliffs.

(3) *Composite mudflows* Periodically masses of decomposed Liassic clay, till, solifluction debris, and peat become mobile and flow with some rapidity down the slopes of the plateau's margins. The civil engineering problems presented by this type of movement are particularly acute along the Antrim Coast Road between Ballygalley Head [D383080] and Glenariff (Hutchinson *et al* 1974).

The Cavan – Down Hill country (18)

From the coast at Donaghadee, Co. Down, southwestwards to Bellananagh, Co. Cavan, is a distance of 150 km, and that line forms the axis of a belt of subdued and scattered uplands. The region is underlain by the southwestward striking rocks of the Longford – Down axis, the Caledonian orogene consisting chiefly of Silurian and Ordovician strata penetrated in the centre by the Newry granite, and overlain in the north by the Trias that floors the Lagan valley and the Comber depression lying between Belfast Lough and Strangford Lough. As we saw earlier (p.13), this region is Ireland's counterpart to the Southern Uplands of Scotland, but the analogy is structural rather than topographical; not only does the Cavan – Down Hill country contain merely a scattering of hills, but those hills rarely exceed a height of 240 m. Some of the hills owe their existence to the presence of a particular structure within the orogene. In Co. Down, for instance, near Newtownards, a Tertiary dolerite sill capping Triassic sandstone forms Scrabo Hill [J478725], and farther to the southwest the indurated aureole rocks of the Newry granite give rise to hills such as Slieve Croob [J317455]. In other cases no such obvious structural explanation is to hand, and groups such as the Holywood Hills to the southeast of Belfast Lough, or the uplands to the northwest of Newtown Hamilton in Co. Armagh, must all be accounted for as merely random residuals.

Over much of the region the uplands are so widely scattered that locally the topography can be described only as undulating lowland. One such lowland forms the Ards peninsula of Co. Down, while another is to be seen around Castleblayney, Co. Monaghan. On all these lowlands there lie thousands of drumlins, the members of the northern Irish drumlin-swarm, and they give rise to the type-example of a 'basket of eggs' landscape — a variety of terrain sometimes known colloquially in Ireland as 'bag of potatoes country'. The drumlins themselves vary in length from some 200 m up to 850 m, and the ill-drained hollows between the drumlins are the site of innumerable lakes, marshes and peat-bogs, although solid rock does frequently crop out in what at first sight appears to be an exclusively drift landscape.

The Tertiary igneous mountains (19)

Between Dundalk and Dublin there lies the major breach in Ireland's upland rim, and commanding that breach in the north there stands a cluster of three compact mountain groups which all share one characteristic in common: they all owe their existence to the presence of intrusive Tertiary rocks. The three groups are the Mourne Mountains, the Slieve Gullion Mountains, and the Carlingford Mountains (fig. 3.13).

The Mourne Mountains are founded upon a Tertiary granite pluton intruded into the Silurian slates and grits of southern Co. Down. In area the pluton covers some 150 km², but it is by no means a simple intrusive body. It consists of five petrologically distinct granites (G1, G2, G3, G4, and G5) successively injected from two centres as a result of ring-dyke formation and cauldron subsidence. Around the pluton the Silurian rocks have experienced some metamorphosis, and this has toughened them somewhat, so that in many places they form foothills fringing the granite massif. In any particular locality the extent of the foothills reflects the width of the metamorphic aureole, and that is in turn related to the steepness with which the granite contact itself is plunging. The granite mountains forming the core of the upland constitute a compact range in which thirteen peaks rise to heights in excess of 600 m. But although it is a compact group, it is usual to divide the Mournes into two parts by means of a line drawn more or less along the road that runs northwards from Kilkeel, through Deers Meadow, and past the Spelga Reservoir. To the east of that line there lie the Eastern, or High Mournes, and to the west there lie the Western, or Plateau Mournes.

The Eastern Mournes are underlain by granites of the G1, G2 and G3 types, and they form a semicircular ridge extending from Slieve Muck in the west to Chimney Rock Mountain in the east, with a central 'stalk' that trails southwards via Slievelamagan to Slieve Bignian. On either side of that 'stalk' there lies a large glaciated valley: the reservoir-floored Silent valley to the west and the forested Annalong valley to the east. The mountains themselves — and all but two of the 600 m peaks lie in this division — are great convex-topped domes which to the south of Newcastle, in the words of the song, really do sweep majestically down to the sea. Dilation of the granite following its unroofing is doubtless chiefly responsible for these immense whale-backed forms, and interestingly that unroofing is not quite complete because a Silurian outlier still survives high on the slopes of Slieve Donard, while on Slieve Muck a tongue of Silurian rocks climbs to the very summit of the mountain there to form a prominent escarpment facing towards the south and east. The Eastern Mournes were affected by vigorous glacial erosion, and particularly striking are the cirques which have been quarried out of the crests of many a granite dome. The activity of glacier-ice must be a partial explanation for the fact that steep slopes are unusually prevalent in the sub-region. Indeed, Colhoun (1967) estimates that 40% of the slopes in the Eastern Mournes exceed an angle of 20°, and that 10% exceed an angle of 35°. The steep granite slopes are in striking evidence when the Eastern Mournes are viewed from the adjacent lowlands, because in the east there are no Silurian foothills; the granite margins plunge almost vertically and the metamorphic aureole is of insufficient width to have much topographical effect.

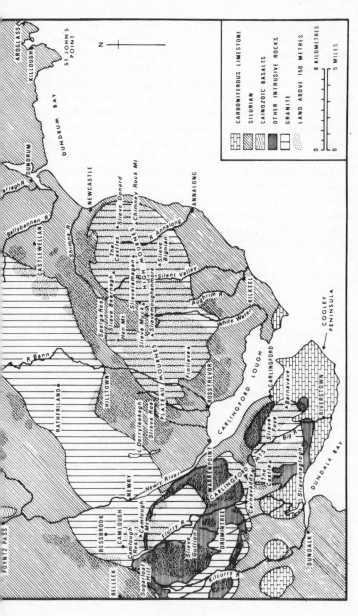

Fig. 3.13   The Tertiary igneous mountains.

To the west of Deers Meadow there lie the Western Mournes, developed mainly in the G4 and G5 granites, and displaying a character very different from that of the Eastern Mournes. The topography here is lower with only two peaks attaining a height of 600 m; there are many smooth uplands which impart to the sub-region the appearance of being a plateau rather than a range of mountains; cirques and other glacial landforms are much less in evidence than in the Eastern Mournes; and steep slopes are much less frequent with only 18% of the slopes exceeding 20° and only 1% exceeding 35° (Colhoun 1967). To some extent the differences between the Eastern and Western Mournes are explicable in terms of geological structure. The granites in the western portion of the pluton never rose as high in the crust as did the granites in the east, and whereas the eastern granites are now exposed to some depth, the western granites must have been laid bare much more recently, and denudation there is still at work in the roof zones of granites G4 and G5. As a result, the granite margins dip very much less steeply in the west than in the east, and the rocks of the Silurian cover therefore feature prominently in the landscapes of the Western Mournes. In the northwest, around Slieve Roe [J184240], there is a broad belt of foothills developed in the aureole zone, and large Silurian outliers survive amidst the outcrop of the G4 granite at Finlieve [J237202] and Slievemoughanmore [J250240].

To the northeast of Rosstrevor the Western Mournes are bisected by a fault-guided depression that extends for 12 km to a point lying just to the east of Hilltown. This is but one of a number of structurally controlled valleys which exist throughout the Mourne region, and Rohleder (1932), in particular, has drawn attention to the many rivers which are orientated either north-north-eastwards or south-south-westwards, thus paralleling one of the granite's major joint systems. There is one other type of structurally controlled feature present in the region, a type of feature which the Mournes display better than any other part of Ireland — tors. They are many and widespread, vary in height from 3 m to 45 m, and are located both on summits and on valley-sides. Good examples are to be seen in the Western Mournes at Hen Mountain [J245275], and in the Eastern Mournes on Slieve Bearnagh [J313280] and at the Castles of Commedagh [J347278], where the tors are formed in the well-jointed G2 granite on the western slopes of Slieve Donard.

To the west of the Mourne Mountains there lies the drumlin-covered

Newry Lowland and the fault-guided course of the Newry River, which in late-glacial times was the outlet for the southward-spilling waters from Lough Neagh (see p.69). Immediately to the west of the Newry River there lies the second group of Tertiary igneous uplands —the Slieve Gullion Mountains. As we have seen, many an Irish upland possesses a close relationship with the underlying geological structures, but nowhere is that relationship more strikingly manifest than in the country to the southwest of Newry. The underlying country-rock is the Caledonian Newry granite and into this there have been intruded a variety of Tertiary igneous rocks in the form of a ring-complex some 10 km in diameter. As the intrusion proceeded, the central block of the complex was perhaps affected by cauldron subsidence, but for whatever reason that central area became the site of alternating beds of granophyre and dolerites, all of them lying more or less horizontal.

Today the granophyres and dolerites survive only at the eye of the complex (some thirteen different beds now exist) and there they form the steep, stepped flanks of Slieve Gullion itself, rising to a height of 583 m. Around this central peak there lies an almost continuous marshy-floored lowland developed in the Newry granite, and then, beyond this lowland, there rises a girdle of rugged hills fashioned from the intrusive rocks of the ring-dyke, these being chiefly tough granophyres and felsites. The whole structure stands out very clearly on the half-inch map, and even the tectonic form of the complex can be read from the topography, because the gaps in the ring of hills are mostly the site of southeastward-striking faults. Where a wrench-fault offsets the intrusive rocks of the ring, as it does at Camlough Reservoir [J030245], then the topography, too, is offset, and in the case just cited, the eastern end of Sugarloaf Hill is displaced 2 km to the north of the western face of Camlough Mountain. One trivial but rather unexpected feature of the region is that some olivine-dolerite sills have been affected by solution weathering so as to form features akin to the lapiés of a karstic landscape (Reynolds 1961). While awaiting customs-examination at Killeen — the border between Northern Ireland and the Republic runs through the complex — the geomorphologist might also like to notice the glacial tail to the south of Slieve Gullion around the settlement of Drumintree.

Just to the southeast of the Slieve Gullion ring-complex, and making contact with it, there lies the final group of Tertiary igneous mountains — the Carlingford Mountains. Here the igneous complex occupies the greater part of the Cooley peninsula lying between Carlingford Lough and Dundalk Bay, and its age is roughly comparable with that of the Slieve Gullion complex. In the Cooley peninsula the local country rocks are Silurian greywackes overlain in the east and south by the Carboniferous limestone, but into these strata there have been intruded a variety of igneous rocks. The igneous history has clearly been complex, but there are perhaps three main stages to be recognised: firstly, the intrusion of a gabbro ring-dyke; secondly, the intrusion of a gabbro laccolith; and finally the intrusion of another ring-dyke, this one consisting of granophyre. On the geological map these rocks appear today as a ring elongated from northwest to southeast, its centre consisting of granophyre and the rim being formed chiefly of a discontinuous outcrop of gabbro.

As at Slieve Gullion, the structure of the complex is closely reflected in the topography. Along the outcrop of the gabbro and the margin of the granophyre, there is developed a rampart of steep-sided, boulder-strewn mountains extending around the ring from Slievenaglogh in the southwest, via The Castle, Carnavaddy, Carlingford Mountain and Slieve Foye, to Barnavave in the southeast. Inside this ring of hills the granophyre has evidently proved rather less resistant to denudation, and the heart of the complex has been eaten out to form Glenmore, although the development of the glen must have been facilitated by the presence of a major fault which strikes northwestwards through the middle of the complex. It is that fault which appears to control the course of the Big River in Glenmore, and at the head of the glen The Windy Gap — it usually lives up to its name — lies on the same line of structural weakness.

The waters of the Big River enter the sea near Riverstown, whence, weather permitting, the observer sees the low-lying coast of the Central Lowland extending southwards for 80 km to the shores of Dublin Bay. There, through the haze over Dublin itself, the outline of the mountains at the northern end of the Leinster chain may be perceived dimly. That vista completes our geomorphic circuit of Ireland.

4 Tertiary history

In 1887 Sir Archibald Geikie, the Director-General of the Geological Survey of Great Britain and Ireland, visited the storm-swept summit of Slieve League, Co. Donegal. There, resting upon the mountain's quartzites, he saw two tiny outliers of what he took to be basal Carboniferous sandstone, and he interpreted their presence atop the mountain, at a height of 607 m, as striking proof of 'the stupendous denudation' experienced by Ireland's Carboniferous strata. With his conclusion there need be no quarrel; great thicknesses of Irish Carboniferous strata have indeed been so denuded. But in one important respect our interpretation must differ from that current in Geikie's day. He and his contemporaries believed that the greater part of Ireland had made its final emergence from the sea late in the Carboniferous, and that the reduction of the Carboniferous strata to their modern form represented the achievement of denudation operating throughout the entire post-Carboniferous interval. This view is no longer tenable. In a paper read at the Dublin meeting of the British Association in 1908 Cole suggested that Ireland had in all probability been submerged beneath the Cretaceous sea, and since Cole's day wide currency has been accorded to the view that Ireland's Palaeozoic rocks were once mantled by representatives of various Mesozoic formations — formations which denudation has now all but completely removed from the Irish scene. The denudation which Geikie termed 'stupendous' thus not only involved formations of whose Irish existence he was largely unaware, but it is a denudation which, far from occupying the entire post-Carboniferous interval, has to be telescoped into the very much shorter time-span represented by the Cainozoic era.

The island's primordial surface

Ireland's emergence from the sea is doubtless an event to be associated with the alpine orogeny, and at the time of its upheaval the new landmass was in all probability veneered with a coat of that Upper Cretaceous chalk which now survives extensively only beneath the basalts of the northeast (Hancock 1961). There, in the northeast, the stratigraphical record permits a crude estimate of the date of the region's final emergence because the basalts, which are both Palaeocene and sub-aerial, rest directly upon the Cretaceous chalk, and the emergence must therefore have taken place somewhere around 70 million years ago. Immediately beneath the basalts there lies a weathered chalk surface carrying a residue of clay-with-flints, and this horizon is a surviving specimen of the type of early Tertiary surface that must have been widespread in Ireland. Outside the northeast, the Mesozoic and Tertiary hiatus in the stratigraphical record renders a dating of the emergence much less certain, although it should be noted that the early to mid-Tertiary deposits at Ballymacadam, Co. Tipperary, are of sub-aerial origin, thus indicating that southern Ireland was perhaps uplifted at around the same time as the northeast.

Aside from the persistence of the chalk in the northeast, much of the evidence for a former chalk mantle lies off the Irish coast. Long ago, during the course of some pioneer investigations of Irish submarine geology, flints-and-chalk fragments were dredged from the continental shelf off Ireland's western seaboard from the latitude of Kerry northwards to that of Donegal (Cole and Crook 1910), and much more recently Cretaceous outliers have been located in the Celtic Sea off southern and southeastern Ireland (Blundell *et al.* 1971, pp.365-70; Eden *et al.* 1971, p.125; Dobson *et al.* 1973, pp.30-1; Naylor and Mounteney 1975, pp.72-3). But by no means all of the evidence lies concealed upon the sea-bed. More than a century ago the officers of the Geological Survey reported the presence of flints-and-chalk fragments in the drifts of southern Co. Waterford (Geological Survey 1861, p.21), and later the same phenomena were discovered in the drifts of both counties Cork and Limerick (Lamplugh *et al.* 1905, p.105; 1907, p.64), all these localities of course lying far removed from any visible Cretaceous outcrop. The burden of the evidence was clear enough, but even so the Irish geological world was hardly prepared for a startling announcement that came in 1960. Walsh had discovered a tiny outlier of Senonian chalk set amidst the Upper Palaeozoic strata at Ballydeenlea 10 km to the north of Killarney (Walsh 1959-60, 1966). Since 1960 that outlier must have been one of the most frequented of all Irish geological sites. The trail of hammer-chipped fragments around the outcrop bears testimony to the geological vandalism of what should surely have been preserved as a site inviolate.

Prior to Walsh's discovery at Ballydeenlea, geomorphologists had grown accustomed to thinking of the former chalk mantle as a kind of roof that once passed over Ireland at a level far above that of the modern topography. In 1951, for example, Linton suggested that the base of the Irish chalk must have stood some 1500 m above present sea-level, while seven years later Whittow (1958) interpreted the form of the ancient chalk surface as a gently folded geanticline possessed of a crest-line extending from Wicklow to Kerry and standing at about 900 m O.D. In short, there had developed the notion that the modern Irish landscape was a result of the steady denudation of a massive, chalk-veneered block that was uplifted during the late Cretaceous, and it was upon the southward-dipping surface of this block that such supposedly superimposed rivers as the Barrow, the Nore and the Shannon were presumed to have taken their origin. There was, of course, need to explain how the surviving chalk in northeastern Ireland came to lie at so low a level, but this difficulty was easily resolved by viewing the entire region as a graben downfaulted some 1000 m or more during the same diastrophism as that which yielded Ulster's legacy of Tertiary igneous rocks. Throughout the remainder of Ireland, however, post-Cretaceous denudation was believed to have proceeded on its block-reducing way without any interference from either vulcanism or earthmovement.

In the light of our present knowledge it is clear that this was too simple an interpretation of Ireland's Tertiary history. It is a view which was first shaken in 1957 when Watts discovered that the pipeclay and lignite at Ballymacadam, Co. Tipperary, are early or middle Tertiary in age. The deposits lie at a height of only 75 m, and while they rest upon limestone and may, therefore, have suffered some solution subsidence, they lend no support whatever to the belief that the early Tertiary surface of southern Ireland stood hundreds of metres higher than the present topography. Of itself this discovery was perhaps hardly sufficient to overturn what by 1957 had become the traditional interpretation of Ireland's geomorphic history, but only three years later there came Walsh's exciting announcement of his discovery of the chalk at Ballydeenlea. This second event had a dual impact. On the one hand it was welcome as affording the first indubitable evidence that the chalk sea had indeed transgressed southern Ireland. But on the other hand it arrived like a cat in the geomorphic dovecot, because the Ballydeenlea chalk lies at a height of a mere 80 to 120 m O.D. and it therefore lies at far too low a level to be reconciled with the traditional concept of a sub-Cretaceous surface sweeping across southern Ireland at a height of more than a thousand metres. At Ballydeenlea, as at Ballymacadam, there has in all probability been some solution subsidence, but the relationship of the Ballydeenlea deposit to the adjacent Namurian beds leads Walsh (1966) to conclude that the floor of the chalk sea in the vicinity must have stood at between 120 and 220 m O.D. Evidence from the nearby Gweestin valley draws Walsh (1965) to a similar conclusion. There he found a group of breccias of presumed Cretaceous or early Tertiary age, all of them lying at a height of aroung 44 m O.D., and all of them seeming to reinforce the geomorphic lessons of Ballymacadam and Ballydeenlea.

The deposits at Ballymacadam, Ballydeenlea and in the Gweestin valley afford the only available stratigraphical evidence directly relevant to a discussion of the level of Ireland's early Tertiary surface. But in northern Ireland there is to hand some less direct stratigraphical evidence which could long ago have been used to throw doubt on the traditional interpretation of the Irish Tertiary (George 1967). In the north the Cretaceous beds rest upon rocks varying in age from the Jurassic near Larne, to the Dalradian schists of northeastern Co. Antrim, and the import of this fact should have been crystal clear: Ireland's Mesozoic and Palaeozoic rocks had undergone extensive denudation *before* the Cretaceous transgression. If the pre-Cretaceous rocks of the north were so extensively removed in advance of the Cretaceous transgression, then is it really conceivable that elsewhere in Ireland there survived until the late Mesozoic courses of pre-Cretaceous strata sufficiently thick to raise the region's chalk roof to the high level demanded by the traditional view? The answer must surely be in the negative. And let it be made quite clear that the chalk itself offers no convenient *deus et machina*; although the chalk lying off the Irish coast may be up to 600 m thick, the onshore chalk seemingly never attained such thicknesses, and today the surviving Irish chalk rarely exceeds a thickness of 100 m (Naylor and Mounteney 1975, p.73). The early Tertiary surface therefore cannot have been raised to a high level by the presence of great thicknesses of Cretaceous strata. Indeed, in some places in the north the chalk was so thin as to have been completely eliminated by denudation before the extrusion of the Palaeocene plateau basalts.

Now we have to face a dilemma. On the one hand the stratigraphical evidence just discussed might seem to indicate first, that Ireland's early Tertiary surface must have stood at a comparatively low level, and second, that the scale of post-Mesozoic denudation might appropriately be described as 'modest'. But on the other hand there is equally convincing geological and geomorphic evidence suggesting that during the post-Mesozoic interval Ireland has been the scene of denudation conducted upon nothing less than the grand scale. To this evidence we must now turn our attention.

Evidence of large-scale Tertiary denudation

This evidence may conveniently be arranged into the three categories that follow.

(1) The early Tertiary plutons of Slieve Gullion, Carlingford, and the Mourne Mountains have clearly all been exposed by the removal of thick masses of cover-rock. The Mourne granite is a particularly interesting case because its emplacement evidently occurred without any doming of the surrounding Silurian strata (Richey 1927), and the difference in height between the summit of the range's most elevated granite peak — Slieve Donard — and the level of the adjoining Silurian lowland thus affords a minimum value for the depth of post-Palaeocene denudation. That value is a figure of not less than 800 m.

(2) The Palaeocene basalts of the northeast have clearly been reduced both in thickness and extent as a result of vigorous denudation. Their thinning is amply demonstrated in Co. Antrim by the existence of numerous outliers of the Upper basalts resting upon the red rocks of the Interbasaltic Horizon, and Charlesworth (1963a, p.372) suggested that in eastern Antrim as much as 450 m of basalt may have fallen victim to denudation. Reduction in the extent of the basalts is proved by the presence of basaltic outliers at Markethill, Co. Armagh, at nearby Poyntz Pass, Co. Down, and, still farther to

the south, amidst the igneous complex of Slieve Gullion. Indeed, George (1967) has suggested that the basalts may once have covered the greater part of northern Ireland, although here a word of caution must be intruded because the Lough Neagh clays contain detrital material obtained from a variety of different rocks, thus indicating that formations other than the basalt must have been exposed at the surface during the later Eocene and the early Oligocene (Fowler and Robbie 1961, pp.128-30).

(3) As we saw in Chapter 1, there exists in northern Ireland a multitude of dykes of presumed Tertiary age. Now many of these were clearly intruded long before the local topography had attained anything like its present form. The great U-shaped trough of The Poisoned Glen near Dunlewy, Co. Donegal, for instance, is essentially a subsequent feature excavated along a line of Tertiary dykes (Dury 1959), while the course of another Tertiary intrusion can be traced running across the axis of the glen and then over one of Errigal's quartzite shoulders at a height of 750 m O.D. The evidence for the glen's post-intrusion age could hardly be more convincing. Similarly, the Tertiary Glenfarne dyke runs along the southern slopes of Thur Mountain, Co. Leitrim, across the lowlands around Belcoo, and then onto the northern flanks of Cuilcagh Mountain where the intrusion finally peters out after having traversed some 14 km of variegated terrain. Comparable dykes cut the Carboniferous limestone of Co. Fermanagh on Magho Mountain to the northwest of Enniskillen, and there Wilson (1964) makes the interesting observation that the degree of metamorphism displayed in the wall-rocks around the dykes suggests that the portions of the intrusions now visible may originally have lain at a depth of some 1800 m. It is difficult to adduce such striking geological evidence of Tertiary denudation from southern Ireland because of the paucity of rocks younger than the Carboniferous. Even in the south, however, the large Tertiary dyke system in the Beara, Iveragh and Dingle peninsulas makes it abundantly clear that as late as the Oligocene the ria coast of the southwest possessed a configuration very different from that known to us today.

Possible Tertiary diastrophism

The evidence is incontrovertible; many areas of Ireland have clearly experienced large-scale Tertiary denudation. But how is this fact to be reconciled with the equally clear evidence from Ballymacadam and Ballydeenlea that the early Tertiary surface of Ireland stood but little above the level of the modern topography? The resolution of this dilemma is perhaps best sought by invoking differential Tertiary earthmovements. Those regions where massive denudation has occurred were perhaps areas of uplift, while the regions where denudation appears to have been minimal were perhaps areas that were brought close to the then prevailing base levels as a result of downwarping. In short, as a consequence of differential earth-movement, the ghost of Ireland's primordial chalk surface may today be a surface possessed of considerable amplitude of relief with 'highs' over areas such as Donegal and Down and 'lows' over such sites as those present at Ballymacadam and Ballydeenlea. This is certainly what has happened in the northeast, the one Irish area where the stratigraphical record is sufficiently complete to elucidate the matter. There, during the Tertiary, the Lough Neagh depression sank to receive the sediments which today form the Lough Neagh clays, while to the east and southeast the Upper basalts were stripped off the Antrim uplands and the Tertiary plutons had their roof-rocks removed.

Any detailed discussion of such postulated and widespread Tertiary diastrophism is quite impossible in the present state of our knowledge, but one general observation is, perhaps, permissible. It may be that the basic pattern of the earthmovements involved on the one hand the uplift of the rocks which today constitute Ireland's highland rim, and on the other hand a compensatory downwarping of the rocks of the Central Lowland. Such a pattern of movements certainly offers an attractive explanation for the fact that over the greater part of the Midlands the pre-Carboniferous rocks lie far below sea-level, obscured from view by their limestone mantle, whereas around the rim of the island the very same pre-Carboniferous strata rise hundreds of metres above the limestone to form Ireland's highest mountain peaks. This issue was broached earlier in chapter 2, and if the thesis now being advanced is valid, then it follows that Ireland's saucer-like form may owe almost as much to differential earthmovement as to differential denudation. Equally, it would follow that those Irish rivers which rise in the Midlands and then flow across the highland rim in order to attain the sea are to be regarded as antecedent rivers rather than being cast in their traditional role as examples of superimposed consequent drainage. But that is a subject to be reserved for the succeeding chapter.

The suggestion that Ireland was the scene of differential Tertiary earth movement is hardly novel; various earlier investigators have proffered the idea by way of explanation of a variety of Irish field phenomena. Hartley (1938) suggested that the Sperrin Mountains may have experienced late Tertiary uplift, for example, while Dury (1959) has speculated that central Donegal may have been raised during the Tertiary, that the Foyle valley may have been down-warped, and that Co. Donegal's Glengesh plateau may have been tilted towards the south. But all this is little better than mere speculation. What real, geological evidence is available to support the view that Ireland experienced widespread Tertiary diastrophism? In the northeast, of course, the evidence is completely satisfying, but elsewhere the post-Carboniferous hiatus in the stratigraphical record renders the argument difficult of proof. Even so we are not quite reduced to the level of clutching at geological straws. The many intrustions that cleave the rocks from Co. Donegal southwards to the peninsulas of counties Kerry and Cork must surely be indicative of former earthmovement, and faulting of supposed Tertiary age is certainly encountered in counties Donegal, Mayo and Galway, and at Kingscourt on the borders of counties Cavan, Meath and Monaghan. Gill (1962, p.63) tentatively suggested that some of the late shear deformation in the Ridge and Valley region might be of Alpine age, while farther to the south there are believed to be gentle fold structures present in the Cretaceous beds on the floor of the Celtic Sea (Eden *et al.* 1971, p.125).

In this context we must return to Walsh's discovery of the chalk at Ballydeenlea, because Walsh (1966) poses an interesting question. If, as the evidence would suggest, the topography over the Ballydeenlea site lay at a mere 120 to 220 m O.D. as early as Senonian times, then is it likely that only a few kilometres to the north and south of Ballydeenlea there stood anticlinal Old Red sandstone mountains — ancestors of the present Slieve Mish Mountains and Macgillycuddy's Reeks — towering 800 m above the bed of the chalk sea? It certainly seems reasonable to suggest that the relative levels of the sandstone and the intervening chalk may well have been changed as a result of post-Mesozoic diastrophism, although Walsh himself admittedly concludes that such differential warping is unlikely to have occurred. To dismiss the idea, however, leaves us to face a lingering litho-logical problem. If the Ballydeenlea site *was* overlooked by sand-stone mountains as early as the Cretaceous, then why is no detrital

Old Red sandstone to be found in the Ballydeenlea chalk? As yet the only inclusions known in the chalk are fragments derived from the local Namurian beds.

The evidence just adduced to support a belief in widespread Irish Tertiary diastrophism can hardly be regarded as conclusive, but there are two areas from which there may be drawn evidence of a somewhat stronger nature: from the mountains on the borders of counties Galway and Mayo in the west, and from the mountains of Co. Wicklow in the east.

In Galway and Mayo, Dewey and McKerrow (1963) have drawn attention to two small outliers of basal Carboniferous sandstone lying at a height of around 640 m on the tableland that forms the summit of the Maumtrasna massif. The outliers both rest unconformably upon the local Ordovician strata, and the two authors suggest first, that the Maumtrasna tableland is really the gently domed and stripped surface of the sub-Carboniferous unconformity, and second, that the plane of the unconformity can be extrapolated from beneath the two outliers to form a surface that is tangent to many of the surrounding peaks (fig. 4.1). Eastwards — and this is the point of real significance — the surface sinks until it is finally lost to view at around 30 m O.D. near the shores of Lough Mask where the Ordovician strata still retain their Carboniferous mantle. There, for instance around Clonbur, the basal member of the Carboniferous series is a sandstone which the two authors regard as identical with the sandstone forming the outliers atop Maumtrasna lying 15 km to the northwest and 610 m above the surface of the Clonbur lowland. Clearly the warped form of the unconformity may be attributed to earthmovements occurring at any time during the long post-Carboniferous interval, but Dewey and McKerrow favour the view that the uplift of the Maumtrasna massif was in all probability a Tertiary event — an uplift which may have taken place as recently as the Pliocene.

The second case — that provided by the mountains of Co. Wicklow — would seem to indicate that recent uplift is by no means a phenomenon confined to the far west. In Wicklow it is upon the granite of the Leinster batholith that our attention must focus. The granite is, of course, Caledonian in age, and it had lost its roof as early as Devonian times. Of this fact there can be no doubt; detrital material derived from the granite is to be found in the Old Red sandstone of counties Waterford and Kilkenny, and in the latter county the sandstone actually overlaps onto the granite. By the early Carboniferous, denudation had proceeded sufficiently far as to allow the limestone to overlap onto the granite along a broad front in Co. Carlow, while farther to the north many blocks of granite were incorporated into the Viséan strata that lie beneath the southern suburbs of Dublin city (Davies 1960a). Now although the Palaeozoic unroofing of the granite is amply demonstrated by the stratigraphical record, it is no less clear that even today, in both counties Dublin and Wicklow, denudation is still at work in the batholith's roof zones. This point was discussed in chapter 3 where we noted that part of the batholith's schist roof still survives in central Wicklow, while in the Dublin Mountains some of the summits are developed in a type of granite characteristic of the batholith's marginal zones. All told it is difficult to quarrel with Jukes's (1862a, p.318) conclusions that in central Wicklow the batholith can have lost a thickness of little more than 300 m of granite. Such a situation would be highly unlikely had the granite lain exposed to denudation throughout the whole of post-Devonian time. Until very recently the granite must surely have possessed a protective mantle of Carboniferous or younger strata. But the granite today rises to a height of almost a thousand metres, and this fact must provoke a question: is it likely that the granite could have retained a protective sedimentary cover at such a level and until very late in the Tertiary, when the topography at Ballymacadam is known to have been reduced to a height of only a few hundred metres as early as the mid-Tertiary? It seems much more reasonable to suppose that the crest-line of the granite lay until recently well below its present level, and that the granite lost its sedimentary cover following a late-Tertiary uplift of the entire Wicklow Mountains area — an uplift that was perhaps related to subsidence in the adjoining Irish Sea basin.

If Ireland's coastal margins really did experience differential Tertiary uplift, then an interesting conjecture perhaps becomes admissible. Is it possible that the uplift was not everywhere simultaneous, and that the degree of fragmentation today present in the various peripheral uplands reflects the interval that has elapsed since the uplands experienced their particular share of the diastrophism?

Fig. 4.1 Contours (in feet) on the reconstructed sub-Carboniferous surface in the Killary Mountains region (after Dewey and McKerrow 1963).

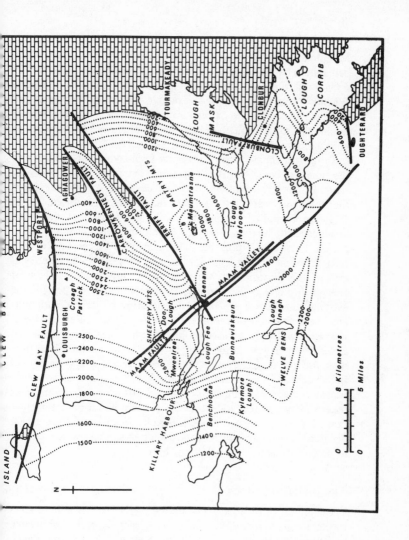

Regions of highly fragmented topography, such as the Killary Mountains, may be areas of relatively early Tertiary uplift, where denudation has had time to cut deeply into the elevated mass. On the other hand, the uplands which retain their cohesion — and here the Wicklow Mountains serve as an example — might be areas of recent uplift where denudation has had insufficient time to effect fragmentation. For the present, however, such matters can be regarded as nothing more than speculation.

So far this chapter has of necessity been somewhat diffuse, and the time has come to offer a tentative summary of the story of Ireland's Tertiary evolution. The island emerged from the sea late in the Mesozoic or early in the Tertiary and it was then surfaced with an extensive mantle of Cretaceous chalk. Denudation of the block commenced under the sub-tropical conditions then prevailing, but in the northeast denudation was soon interrupted by the extrusion of the Palaeocene basalts. That igneous episode was but one manifestation of the diastrophism which probably affected the whole of Ireland throughout the Tertiary, and the cumulative effect of the earthmovements was seemingly the elevation of Ireland's margins and the depression of the interior of the island. In those down-warped areas the Carboniferous limestone has been widely preserved and Tertiary sedimentation resulted in the accumulation of deposits such as the Lough Neagh clays and the clays and lignite at Ballymacadam. The uplifted margins, on the other hand, experienced the vigorous denudation which has left the pre-Carboniferous rocks widely exposed, and which has brought about the fragmentation of so many of the modern peripheral uplands. During that denudation there were formed the planation surfaces which are still to be seen in many of the Irish uplands.

The planation surfaces

The evolution of the Irish landscape during the last few million years is commonly interpreted in terms of various cycles of erosion related to falling Plio-Pleistocene base-levels. Such a view is hardly novel, but in Ireland there is no known sedimentary evidence of the postulated higher sea-levels; the story, such as it is, has to rest solely upon the evidence afforded by planation surfaces which have been varyingly regarded as the result either of direct wave-attack or of sub-aerial denudation related to the conjectural ancient base-levels.

Clearly any conclusions based solely upon such erosional evidence can only be tentative, and in Ireland the problem is further compounded because the fragmentation of the uplands has reduced many of the planation surfaces to remnants so small as to make any interpretation extremely hazardous. Glaciation, too, has taken its toll. In uplands such as those of Kerry and Galway Pleistocene processes have all but obliterated the planation surfaces which perhaps once existed there, while at lower levels, over the entire country, the heavy drift mantle makes difficult the study of the solid rock morphology. In the current state of our knowledge the presentation of an Irish denudation chronology is impossible, and an effort to arrive at such a chronology through statistical analysis proved less than conclusive (Davies 1958). All that will be attempted here is a discussion of the planation surfaces on an essentially provincial basis.

Planation surfaces in Munster

In many parts of Ireland planation surfaces may be poorly developed, but in counties Waterford and Cork there is a display of planation surfaces unique in Ireland and even noteworthy by international standards. Throughout the eastern and southern portions of the Ridge and Valley Province the anticlinal ridges — be they developed in the Old Red sandstone or in the Cork Beds — are topped by remarkably smooth upland surfaces broken here and there by steep-sided glens. The surfaces are much in evidence on either side of the Blackwater estuary (see fig. 3.5, p.33), along the crest of the Great Island anticline, and in the Somerville and Ross country of West Cork, but such mention of specific localities is really superfluous; the surfaces are to be seen almost anywhere between Tramore Bay and Roaringwater Bay — a distance of 180 km. Why this part of Ireland should offer so remarkable an exhibition of planation surfaces is a mystery. Surfaces cut into the region's resistant Old Red sandstone might be expected to prove very durable, but in Co. Cork, around Kinsale and Clonakilty, for example, the surfaces are just as well developed in the Cork Beds which seem to possess no exceptional powers of resistance to denudation.

The planation surfaces of Waterford and Cork early attracted notice (Jukes *et al.* 1861, p.8; Lamplugh *et al.* 1905, p.108; Hallissy

1923, pp.66-9), but it was from Miller (1939, 1955) that the surfaces received their first modern attention, although to present-day eyes his study might seem a mere reconnaissance survey. Miller interpreted the region's denudational history in terms of two major cycles. Firstly, he recognised a South Ireland peneplain stage during which large areas were submerged by a transgressive (Plio-Pleistocene?) sea, its shoreline standing at about 246 m O.D. The strandline of this transgression he traced from the Irish Sea north of Cahore Point, Co. Wexford, along the southern margins of the Blackstairs, Comeragh, Monavullagh, Knockmealdown, Nagles and Boggeragh mountains, to the sea near Skibbereen (see fig. 3.5, p.33), the strandline sometimes being marked, as on the Knockmealdowns, by what Miller regarded as degraded marine cliffs. According to Miller, wave abrasion at this level produced the widespread 'flats' which today stand at between 185 and 246 m O.D. — the remnants of the South Ireland peneplain — the finest example of such a 'flat' being the remarkable bench on the southern flanks of the Knock-mealdown Mountains. From the 246-m shoreline Miller believed the sea to have retreated until a new stillstand was established at a height of 123 m O.D., that being what he termed the Coastal peneplain stage. The strandline of this stage he found less easy of detection, but he regarded the widespread 'flats' in the coastal margins of counties Waterford and Cork as surviving fragments of a wave-abraded Coastal peneplain which today stands at between 60 and 123 m O.D. These fragments — their surfaces often ill-drained — are strikingly evident to the west of Tramore, at Ballyvoyle Head, west of Ballycotton, and around Cork Harbour, and in these localities, as elsewhere, the 'flats' end abruptly at the coast in steep cliffs some 60 m high. Following the Coastal peneplain stage, Miller claimed, the sea-level again sank until eventually it came to rest in its present position.

A few geologists have adopted a very different interpretation from that advanced by Miller; they have argued that the planation surfaces of Waterford and Cork are really exhumed features, being part of the region's sub-Cretaceous floor (Nevill 1963, pp.237-8). But such views have received little support, and most of Miller's geomorphic contemporaries found his general schema convincing. Indeed, during the 1950s attention came to focus not upon the question of whether the South Ireland and Coastal peneplains were real entities, but rather upon the issue of the origin of the two surfaces. Were they really wave-abrasion platforms as Miller claimed, or were they essentially the product of sub-aerial processes? The surfaces, of course, carry no known marine deposits, and Farrington (1951, 1953a, 1961) was able to offer some telling objections to Miller's marine hypothesis, but throughout the debate there was little first-hand contact with the surfaces and cartographic analysis took the place of field investigation. Even a cursory examination of the region is sufficient to suggest that its physiographic history may have been rather more complex than Miller's view would allow, because within the broad height ranges of both the South Ireland peneplain and the Coastal peneplain there exist many individual 'flats' which perhaps deserve to be viewed as representing a number of independent stages. This is certainly the lesson of Orme's (1964) study of the planation surfaces in the Drum Hills of Co. Waterford. There he described fragments of a platform seemingly cut by a transgressive (Calabrian?) sea with a strandline at about 215 m O.D., and below this level he saw evidence of marine benching by a recessive sea at the following levels: 185; 160; 150; 121; 109; 91; 79; and 64 m. Certainly Miller's chronology is insufficiently well grounded to allow its extrapolation to areas outside Munster; claims to have located analogues to the South Ireland and Coastal peneplains in other parts of Ireland must be treated with reserve.

The only other part of Munster where planation surfaces have received detailed study is in the Carboniferous rocks of the Clare Plateau (p.44). There Sweeting (1955) has recorded the existence of surfaces at the heights of 246-62; 185-200; 132-42; 108-18; 71-6; and 46-60 m. The surfaces near the coast she suggested might be of wave-cut origin while those farther inland she preferred to regard as essentially sub-aerial. The identification of cyclical plantation surfaces in an area of almost horizontal strata is nevertheless an exercise of some difficulty and Drew (1975) has doubted the significance of at least some of Sweeting's results.

Planation surfaces in Leinster

The gently rounded summits of the Leinster Mountains have been regarded as preserving the relics of an ancient planation surface standing at over 700 m O.D. The smooth skylines of peaks such as Lugnaquillia or Turlough Hill [T068986] — the latter sufficiently extensive to be the site of the upper reservoir for a pumped-storage power station — are certainly striking, but they probably owe far more to the sheet-jointing of the granite than they do to former base-levels. Such structural control, however, can hardly be invoked in explanation of the rolling granite upland that is traversed by the Military Road from Glencree southwards to Glenmacnass, the upland standing at between 375 and 550 m O.D. This surface has every appearance of being related to some ancient, elevated base-level, and its survival is perhaps a result of the resistant schist of the aureole having provided a local base-level for the rivers which drain the surface towards the southeast.

At lower levels in Leinster the influence of former base-levels is much more positively displayed. In the south the rivers Nore, Barrow and Slaney are all impressively incised into planation surfaces of presumed Plio-Pleistocene age. The incision of the Nore is well seen around Inistioge, and that of the Barrow around Graiguenamanagh where, between Brandon Hill and the Blackstairs Mountains, the river flows in a gorge carved below the ill-drained floor of a surface lying at between 50 and 100 m O.D. Equally striking is the case of Slaney. Above Kilcarry Bridge [S894625] the Slaney is encased in a gorge cut into a remarkably subdued granite plain standing at around 65 m O.D. and forming a portion of the Tullow Lowland. Above the surface there rise a few steep-sided residuals such as Ballon Hill [S826661].

Farther to the north, around Woodenbridge in Co. Wicklow, there is further striking evidence of changing base-levels where the vales of Aughrim, Avoca and Arklow, and the valleys of the eastern and western Gold Mine rivers are all sharply incised into a rolling upland standing at between 60 and 120 m O.D. The same wave of rejuvenation can be traced from the Vale of Avoca, upstream past Rathdrum (the road outside the town's Catholic church affords an interesting viewpoint at T188886), along the Vale of Clara, and into such of the Wicklow glens as Glenmalur, Glendalough, Glendasan and Glenmacnass (see fig. 3.3, p.23). As was suggested earlier (p.28), the trough-end features at the head of most of the glens may in part be the result of drainage rejuvenation. Some 10 km to the north of the Vale of Clara there lies the Roundwood Basin developed in the Bray Group rocks and elongated along their northeastward trending strike. Today, as a result of glacial diversion, the basin is drained southeastwards through the Devil's Glen, but it seems that originally the basin was drained southwestwards along the Caledonian strike and into the Vale of Clara via Laragh. Certainly the floor of the basin bears traces of successive waves of rejuvenation working headwards from the southwest. Five planation surfaces of sub-aerial origin have been identified on the floor of the basin (Davies 1966); the Moneystown (170-200 m), Carriggower (232-44 m), Glasnamullen (248-70 m), Calary (290-6 m), and Ballyremon (310-25 m) surfaces. The Great Sugar Loaf is a quartzite residual rising above the boggy ground of the Calary surface.

In the area between the Leinster Mountains and the sea, Miller (1938) suggested the existence of a much denuded 185-246-m surface — the analogue of Munster's 'South Ireland peneplain' — preserved on the accordant summits of hills in eastern Wexford and Wicklow. On the seaward side of these residuals he believed there could be discerned the fragments of a younger, 60-123-m surface (the 'Coastal peneplain'?) similarly preserved on scattered hill-tops. He regarded the bold Tara Hill [T205623] as a former cliffed island rising above the wave-abraded surface of the lower of these two platforms. But Miller's interpretation was based essentially upon cartographic interpretation; the evidence is tenuous and again his conclusions must be treated with caution. More secure is Farrington's (1929) study of the Dublin region. Using evidence from boreholes, he traced the form of the rock-head beneath the drift of Dublin and its environs, and concluded that there was present a buried planation surface varying in height from 60 to 105 m O.D. Into this surface the Liffey is incised.

Planation surfaces in Connacht

The fragmentation and heavy glaciation of the uplands of Connacht have resulted in the survival of few planation surfaces. Dewey and McKerrow (1963) have interpreted the even skyline of the Maumtrasna massif at a height of around 640 m as a stripped sub-Carboniferous unconformity (see p.87 and fig. 4.1, p.89), while in the Corraun peninsula (see fig. 3.8, p.53) Flatrès (1954, 1957) has described an exhumed sub-Devonian surface standing at around 330 m O.D. At a lower level, Burke (1957) regarded the Iar-Connacht Lowland (pp.46, 195) as a stripped and glacially modified sub-Carboniferous surface cut across the Galway granite. He suggested that earthmovements had given the surface its gentle southward dip and that the straight coastline between Galway and Inveran was a result of the tilted surface meeting the waters of Galway Bay. Alternatively the surface of the Iar-Connacht Lowland might be related to a late Tertiary base-level, perhaps being comparable in age to the peat-engulfed surfaces standing at around 30 m O.D. south of Louisburgh (see fig. 3.7, p.47) and in northern and northwestern Co. Mayo (see fig. 3.8, p.53). In northern Co. Mayo, around Belderg and Rathlackan, this surface ends abruptly at the coast in a series of cliffs which are reminiscent of the termination of the 'Coastal peneplain' along the Munster coast.

Planation surfaces in Ulster

The literature contains many references to the presence of frag-mented planation surfaces in northern Ireland, but no regional synthesis has yet obtained wide currency (Reffay 1972, ch. 4). In Co. Donegal, Dury (1957, 1959, 1964) has referred to a possible 480-m summit-surface in the Slieve League peninsula and elsewhere; a surface at 240-70 m in central Donegal and the Slieve League peninsula; a platform at around 180 m near Lough Derg; some possibly wave-trimmed benches standing at 120-50 m to the north of Killybegs; and various lower platforms, some of them perhaps related to a base-level at 60 m O.D. In the Sperrin Mountains, Thorp (1962) suggested the existence of a staircase of eight unwarped surfaces lying between 150 and 600 m O.D., while farther to the south, in Co. Tyrone, the plain of The Six Towns, lying between Omagh and Draperstown, is often regarded as a planation surface standing at between 180 and 240 m. In the northeast the stepped character of much of the terrain arises from the development of structural platforms on the gently dipping lava-flows, although, in the broader view, the rolling topography of the Antrim Plateau is commonly interpreted as a planation surface standing at heights of between 300 and 380 m. The Antrim glens are incised into this surface and such features as Agnews Hill [D328015], Slemish Mountain [D221053], and Trostan [D179236] are seemingly residual peaks rising above the surface. In the Mourne Mountains, Proudfoot (1954) suggested the existence of an ill-defined summit-surface at 600 m, below which he identified fragmentary planation surfaces — some of them possibly of marine origin, others possibly sub-aerial — at the following metre levels: 92-123; 136-54; 169-200; 228-62;

'302-64; 375-85; 394-480; and 523-54. Finally, at lower levels in Ulster, the drumlin-strewn lowlands of Co. Down, on either side of Strangford Lough, are commonly viewed as a planation surface standing at around 60 m O.D., and the same surface is perhaps represented by the lowlands that fringe Co. Donegal.

The Midlands

The Central Lowland is bevelled across a variety of geological formations and structures, and beneath the drift overburden there seems to be an extensive planation surface lying at heights varying from sea-level up to about 120 m. The surface is perhaps now tilted with a dip towards the west because such east-coast rivers as the Liffey and the Boyne are incised some 30 m below the level of the surface, whereas in the west the surface dips gradually beneath the waters of both Galway Bay and Clew Bay without any geomorphic discontinuity (Corbel 1957, pp.343-4). Close (1867) suggested that the surface might owe its form to the waves of some bygone transgressive sea, and Miller (1939) tentatively equated the surface with the wave-trimmed 'Coastal Peneplain' of Munster. It now seems more likely that the greater part of the lowland has resulted from the solution planation of the underlying limestone in relation to a hydrological base slightly higher than that obtaining today. Such solution planation certainly seems to be the origin of that remarkably level portion of the lowland to the southeast of Galway Bay and known as the Gort Lowland. Elsewhere in the Midlands the drift mantle may well be obscuring a considerable diversity of limestone terrain and there was mentioned earlier (p.22) Murphy's (1962) suggestion that solution phenomena may be widespread in the eastern Midlands.

The limestone underlying the greater part of the Midlands appears to weather rapidly and the surface of the Central Lowland is therefore unlikely to be of any great antiquity. Perhaps the history of the region has followed something like the following sequence: (1) The Carboniferous strata — especially the limestone — were widely exposed as a result of Tertiary denudation. (2) Tertiary diastrophism gave Ireland its saucer-like form by downwarping the Midlands so as to place most of the Carboniferous beds below Tertiary base-levels. (3) In the depressed region there accumulated a considerable depth of Tertiary weathering residues derived largely from the adjacent uplands. (4) A late (Plio-Pleistocene?) fall of base-level allowed the widespread removal of the Tertiary regolith and permitted solution planation to bevel a new surface across the limestone. During this stage a remnant of the regolith was lowered several tens of metres to become the Upper Eocene (?) deposit which now lies at Ballymacadam. (5) The newly formed planation surface was gently tilted and given its westward dip as Ireland's eastern margins experienced their most recent uplift.

Low Plio-Pleistocene sea-levels

All around the Irish coast there is evidence of river valleys having once been related to base-levels much lower than those of the present — valleys which have been partially drowned by a relatively recent rise of sea-level. In the southeast, for example, the rivers Barrow and Nore are both tidal to points far upstream: the Barrow to St Mullin's and the Nore to Inistioge. More decisive is the evidence afforded by the presence of submerged, drift-filled valleys at many localities. Borings at the head of Lough Foyle have shown the rock-head to lie some 30 m below Ordnance datum (McMillan 1957); at Belfast, deep, drift-choked gorges cut into the Trias have been located more than 60 m below the waters of Belfast Lough (Manning *et al.* 1970, pp.123-4); and incised into the Carboniferous strata flooring Dublin Bay there can be traced a former course of the River Liffey lying at a depth of -39 m O.D. (Naylor 1965). Borings in the valley of the Slaney at Wexford, and in that of the Blackwater at Youghal, suggested to Orme (1964, 1966) the presence of a buried valley-in-valley topography related to base-levels at -22 and -45 m O.D., and offshore, between Minè Head and Knockadoon Head, he postulated the existence of a submerged planation surface related to the -22-m base-level. Westward of Knockadoon Head, near Midleton in the Castlemartyr syncline, Murphy (1966) has employed geophysical methods to locate solution cavities in the limestone at depths of up to 180 m and he observes that the weathering responsible must have occurred at a time when the sea stood more than 150 m below its present level. Still farther to the west, at Carrigrohane [W615715] just above Cork City, the rock-head in the Lee valley lies at -24 m O.D., and, assuming a fall of two metres per kilometre, this would place the rock-floor of the Lee's valley at Roche's Point [W826601], to the south of Cork Harbour, at a depth of -78 m O.D. (Lamplugh *et al.* 1905, pp.75-6). Finally, there are the rias of the southwest. There seems much sense in Wright's (Wright *et al.* 1927, p.6) conclusion that the Carboniferous rocks must have been removed from the synclines at the western end of the Ridge and Valley Province during a time of low Plio-Pleistocene sea-levels.

5 Drainage patterns

Even the simplest map is sufficient to demonstrate the puzzling character of the rivers that drain a large proportion of Ireland's surface. Such major rivers as the Slaney, the Barrow, the Nore, the Suir and the Shannon all flow predominantly towards the south, but the problem of the Irish rivers involves much more than mere parallelism and a southward orientation. All the rivers just mentioned compound the enigma by flowing sluggishly for long distances over the plains of the Irish Midlands (in every case except the Slaney they flow over the Carboniferous limestone of the Central Lowland) before entering constricted valleys which (in every case save that of the Shannon) carry the rivers across the pre-Carboniferous rocks of Ireland's upland rim. In short, the rivers display little or no relationship either to the topography or to the underlying geology. The Slaney (see fig. 3.3, p.23) flows across the granite of the Tullow Lowland (see p.24) and then passes on its way to Wexford through the Bunclody Gap which breaks the continuity of the southern part of the Leinster Mountains, leaving the Tinahely Hills and the Cummer Vale ridge to the northeast, while to the southwest there rears the impressive mass of the Blackstairs Mountains. The Barrow and the Nore both flow across the limestone underlying the south-eastern portion of the Central Lowland in order to reach the south coast, the Barrow using the Graiguenamanagh Gap between the Blackstairs Mountains and Brandon Hill, while the Nore traverses the Inistioge Gap between Brandon Hill and the Curraghmore plateau (see fig. 3.3, p.23). These three gaps are all well displayed on Half-Inch Sheet 19. The Suir flows southwards over the limestone of the Central Lowland via Templemore, Thurles and Caher, but where it enters the Ridge and Valley Province the river swings to the east into a strike course which follows the limestone floor of the Carrick Syncline. The ribbon of limestone, however, dies out before the Suir reaches the sea, and in its lower portion the river traverses the Ordovician rocks of the hills and planation surfaces around Waterford Harbour.

The Shannon differs from the rivers farther to the east in that it does not flow across the pre-Carboniferous rocks of the upland rim, but the river does intersect one of the Southern Mountain inliers in a manner that fully entitles the Shannon to be classed with the discordant rivers already mentioned. The Shannon rises only 40 km from Sligo Bay and it flows sluggishly southwards over the limestone of the Central Lowland for a distance of some 220 km (see p.22 and fig. 3.2, p.19). To the south of Lough Derg there rises across the path of the river the resistant Silurian and Old Red sandstone inlier that forms the Slieve Bernagh and Arra Mountains. This obstacle the river could largely have circumvented; it might have flowed over the limestone plain either to the east or to the west of the inlier, and had it done so the Shannon could have maintained its close association with Carboniferous strata from its source right down to the sea. But the river does otherwise. Below Lough Derg it bisects the inlier via the striking Killaloe Gorge, leaving the Arra Mountains to the east and the Slieve Bernagh to the west (Kilroe 1907b; Farrington 1968).

The puzzling behaviour of these major Irish rivers attracted the attention of Jukes as far back as 1862 (Jukes 1862b) and in a now classic paper he suggested that all the southward-flowing rivers of southern Ireland had originated as consequent rivers initiated upon a southward-dipping surface following Ireland's final emergence from the sea. These consequent rivers, he suggested, were then superimposed onto the pre-Carboniferous rocks as the younger strata were worn away, thus yielding the present discordant drainage pattern. Over the hundred years that followed, Jukes's general thesis obtained universal currency, and by the 1950s it was generally accepted that the southern Irish drainage had originated when a Cretaceous-mantled and southward-dipping block emerged from the sea late in the Mesozoic (Charlesworth 1953, pp.188-93).

Widely though it has been accepted, Jukes's superimposition hypothesis has to face some serious objections, the four chief of these being as follows.

(1) Much of the south of Ireland is covered with Carboniferous

limestone, and during the Tertiary, before denudation had reduced the limestone to its present limits, the outcrop of that particular series must have been even more extensive. Under such conditions karstic landscapes must have been widely developed, and during this interlude the direction of the subterranean drainage-lines would be controlled solely by the local structure of the limestone. It is difficult to see how any ancient north-to-south consequent rivers could have survived such a karstic episode.

(2) Jukes and his followers believed that during post-Mesozoic time Ireland, apart from northeastern Ulster, had experienced a period of tranquility free from the effects of diastrophism — a period during which the ancient rivers were gradually superimposed. But, as we saw in the previous chapter, there is increasing evidence that the impact of Tertiary disturbance was felt far beyond the confines of Ulster. Therefore, there must now be admitted the possibility that Ireland's ancient consequent drainage — whatever its form may have been — was deranged by Tertiary earthmovements.

(3) Again harking back to chapter 4, the stratigraphical evidence now available throws doubt on the notion that Ireland's original chalk surface was a plain standing higher than the summits of the present mountains. If there was no such high-level surface, then the entire superimposition hypothesis collapses.

(4) Study of some of the supposedly superimposed rivers reveals them to possess a history more complex than the superimposition hypothesis would allow. The long north-to-south reach of the Barrow, for instance, between Athy and Goresbridge, is essentially a subsequent reach where the river follows the narrow limestone outcrop lying between the granite of the Tullow Lowland to the east and the Namurian beds of the Castlecomer plateau to the west. Equally, it now seems that in its present form the course of the Shannon through the Killaloe Gorge is the result of glacial diversion rather than superimposition (Farrington 1968).

For these reasons, and despite its long pedigree, the superimposition hypothesis must be viewed with some scepticism. In future discussions of the Irish drainage pattern two factors must surely receive far more consideration than they have been accorded hitherto. Firstly, there should be borne in mind the possibility that Tertiary diastrophism may have led both to the existence of some antecedent drainage and, through differential uplift, to the initiation of other rivers *de novo*.

Secondly, much more attention needs to be paid to the possibility of the glacial modification of pre-existing drainage patterns. If the present Killaloe Gorge on the Shannon is a result of ice action, then what other drainage anomalies may have been brought into existence by the same agent? For instance, could the gaps at Bunclody, Graiguenamanagh and Inistioge owe something to the activity of glacial meltwaters? Certainly during the last glacial episode, when ice occupied the Irish Midlands (see fig. 6.2, p.121), drainage had little alternative but to move southwards in the direction now followed by the Slaney, the Barrow and the Nore.

The rivers of Munster

Over the greater part of the Ridge and Valley Province the drainage closely reflects the geological structure, and the trunk streams are eastward-flowing strike rivers occupying the synclinal valleys (see fig. 3.5, p.33). On the smaller scale, too, there is adaptation to structure, because in those synclines where the limestone still survives extensively it usually forms slightly higher ground than the Lower Limestone shales, and there is a tendency for the subsequent rivers to follow the shale outcrop at the margin of the synclines (see p.35). Alongside such examples of close adaptation to the local geology, however, the region displays some remarkable cases of rivers which have a total disregard for the underlying structures. The most striking illustrations of such discordance are provided by the Blackwater and the Lee. Both rivers follow synclines eastwards to within a few kilometres of the sea and then they turn abruptly southwards, forsaking synclinal routes to the sea in favour of more difficult courses across the neighbouring sandstone anticlines. The Blackwater is the case *par excellence*. It flows along the Carboniferous-floored North Cork/Dungarvan syncline for 60 km and by Cappoquin the river is within 18 km of the sea at Dungarvan and at a height of only 6 m O.D. The synclinal valley continues between Cappoquin and Dungarvan, and despite the presence of a certain amount of drift, the floor of the valley lies almost entirely below 30 m O.D. But the Blackwater does not follow this obvious route to the sea. At Cappoquin it swings through a right-angle and makes its way southwards to the sea at Youghal — a course that takes the river across the uplands developed upon no less than four sandstone anticlines. Each of these uplands has a crest-line standing at least

100 m above the level of the Blackwater at Cappoquin, and the steep-sided gorges which the river has cut across the sandstone ridges afford a striking contrast to the wide synclinal valley that lies before the river at Cappoquin.

The Lee below Cork city displays the same anomalous pattern. There, prior to the drowning which formed Cork Harbour, the river turned southwards out of the Cork syncline to enter transverse gorges cut across sandstone uplands developed on both the Great Island and Ballycotton anticlines. Still farther south in Co. Cork the Bandon River is usually regarded as a third river of the same genre. The Bandon flows eastwards along a syncline in the Cork Beds for 32 km, but near Inishannon the river turns abruptly to the southeast to attain the sea at Kinsale Harbour.

Jukes (1862b) was the first to attempt an explanation of these Munster drainage anomalies, and, although modified in some details, his basic thesis has for long been generally accepted (Miller 1939; Charlesworth 1963a, pp.437-8). According to this now traditional interpretation, Munster was originally crossed by some of those southward-flowing consequent rivers supposed to have been initiated upon the chalk surface following Ireland's final emergence. Elsewhere, as we have just seen, some of the consequent rivers were believed to have survived in a superimposed form to become such modern rivers as the Shannon and the Suir, but in the Ridge and Valley Province the original consequent pattern was supposed to have been almost completely obliterated by the development of major subsequent rivers along the Armorican structures. In a few places, however, fragments of the old southward-flowing consequent rivers were held to have survived to become the discordant north-to-south reaches of the Blackwater, Lee and Bandon, while the subsequent reaches of the three rivers were regarded as overgrown, right-bank tributaries which had extended themselves along the strike at the expense of neighbouring and more westerly consequent rivers. The absence of any westward-flowing, left-bank subsequent tributaries was usually explained by the suggestion that the westward rise of the synclinal axes had encouraged all the strike rivers to flow in an easterly direction.

The objections to any hypothesis which derives the modern Irish rivers from a postulated high-level chalk surface have already been outlined, and in view of these objections the traditional interpretation of the Munster drainage must surely be called into question. The problem seems to demand a fresh approach, and in particular it seems useful to focus attention upon ways in which the Munster drainage anomalies might have developed during or after Tertiary karstic phase and quite independently of the drainage elsewhere in Ireland. Explanations involving the glacial diversion of drainage here seem to be of little assistance (Lamplugh *et al.* 1905, pp.6-7) and the hypothesis that follows has been offered as a possible fresh solution to the problem (Davies and Whittow 1975). The hypothesis relates only to those areas of Cork and Waterford which formerly possessed a Carboniferous limestone cover (i.e., to the area north-ward of the Cork Harbour to Kenmare line) and it envisages three stages in the evolution of the drainage pattern.

Stage I: In the mid-Tertiary the surface of the region stood many metres above its present level and the Old Red sandstone was widely obscured beneath the Carboniferous cover, the landscape being developed chiefly in the Carboniferous limestone. To the north the limestone surface was overlooked by an arcuate belt of hills developed upon the resistant sandstone anticlines and extending from the Comeraghs in the east, through the Knockmealdowns, the Nagles and the Boggeragh mountains, to the Shehy Mountains in the west. To the south of the limestone country lay the province of the Cork Beds, but the nature of the topography there is neither clear nor significant. The limestone itself probably formed a karstic area broken only by the occasional inlier of anticlinal Old Red sandstone, and it is the form of those inliers which must now claim our attention.

According to this hypothesis, all the region's synclines were originally long, narrow structures which tapered both westwards *and* eastwards. The Rathcormack syncline (see fig. 3.5, p.33), although complicated by faulting at its eastern end, still displays something of this form, as in miniature do those many surviving Carboniferous lenses which are set into the sandstone and which mark the former position of major synclinal outliers of Carboniferous strata (see, for example, the Blarney syncline). In particular, it is assumed that the two major synclines which are today open to the sea in the east — the Dungarvan and Cork synclines — were originally closed off by the eastward convergence of the bounding sandstone anticlines. Associated with this tapering of the synclines, the hypothesis assumes further that where only a narrow anticline separates two synclines, then the axis of the anticline will sag parallel to the axes of the synclines. Thus, anticlines will be lowest where the corresponding synclines are widest in the same way that the freeboard of a sheered ship is lowest where its beam is greatest. In the absence of any generally recognised term for such structures, they may perhaps be termed 'boat-shaped synclines'.

Returning now to the sandstone inliers, these would first appear at the eastern and western ends of the boat-shaped synclines where the anticlinal axes reached their highest level. The rivers draining off the inliers, and off the sandstone uplands to the north, all passed underground as soon as they crossed onto the limestone. Any earlier drainage directions had been obliterated during the karstification, and within the limestone the drainage followed only structural lines.

Stage II: Continued denudation of the Carboniferous strata caused the sandstone outcrops to increase in extent as they were exhumed, each outcrop forming an upland because of the sandstone's greater resistance to denudation. As the Carboniferous rocks were peeled away, the streams generated upon the sandstone uplands progressively extended themselves in a downstream direction across newly exposed surfaces that were essentially the stripped Devonian/Carboniferous interface. These streams draining the sandstone uplands flowed predominantly towards the south, perhaps because the anticlines — and thus the Devonian/Carboniferous interfaces — were asymmetrical, with steep limbs on the north and more gently sloping limbs to the south (see p.32). This preferred southward orientation of the dip-drainage remains strikingly evident down to the present day.

In those localities where the Old Red sandstone anticlines reached their highest level, the Carboniferous rocks were soon reduced to a series of long and narrow east-to-west synclinal outliers, each outlier almost completely surrounded by anticlinal sandstone ridges. Subterranean drainage from such isolated outliers was impossible; they became waterlogged and developed a surface drainage. The confinement of the Carboniferous strata to such outliers is regarded as marking the close of the karstic phase in that particular locality. The waters draining off the surrounding sandstone, and onto the discrete synclinal outliers, clearly had to escape from the waterlogged basins and they did so at the point where the flanking anticlines were lowest. The location of the outlet streams must in each case have been determined by the position of the last surviving 'bridge' of Carboniferous rocks across the adjacent sandstone anticlines, and normally this 'bridge' would lie about half way along the length of a boat-shaped structure, where the syncline was widest and its flanking anticlines were lowest. The 'bridges', of course, soon disappeared as a result of continued denudation and the outlet rivers then became superimposed onto the sandstone anticlines beneath. It is suggested that the modern discordant north-to-south elements in the drainage of Waterford and eastern Cork originated in this manner.

Stage III: In the Dungarvan, Ardmore, Cork and Cloyne synclines either denudation, or more probably faulting, has removed the sandstone from the eastern portion of the synclines. This removal has opened the synclines to the sea at their eastern end and eliminated the westerly flowing rivers that once drained down towards the outlet river near the median point of the syncline. Eventually the Blackwater must surely abandon its discordant course below Cappoquin in favour of the synclinal route to the sea, and the fact that this has not yet happened must indicate that the opening of the Dungarvan syncline at its eastern end was, geologically speaking, a very recent event.

Davies and Whittow have applied this hypothesis to the detailed development of both the Blackwater and the Lee; whether it can also be applied to account for the course of the Bandon must remain in doubt because that river flows entirely over the Cork Beds in a region where the Carboniferous limestone seemingly never existed. If the hypothesis is valid, however, the discordant north-to-south elements in the drainage are not, as formerly supposed, the oldest rivers in the region but rather the youngest. Similarly, the strike rivers are not subsequents developed by headward extension, but rather a form of longitudinal consequent developed in response to the topographical dips revealed upon the flanks of the sandstone anticlines as the Carboniferous rocks were peeled away.

Rivers in the uplands of Leinster

In Leinster there are essentially three types of river. Firstly, there are the discordant rivers which cross the Caledonian axis of the Leinster Mountains — the Slaney, the Barrow and the Nore. Secondly, there are shorter rivers which are also discordant in relation to the Caledonian structures, and which drain either northwestwards or southeastwards off the flanks of the Dublin and Wicklow mountains. Finally, there are a number of strike streams picking out the region's Caledonian grain and flowing either northeastwards or southwestwards.

The anomalous courses of the Slaney, the Barrow and the Nore have already been discussed, and this first category of stream need detain us no further. The rivers in the second category include among their number the northwestward flowing Liffey and Kings River on the northwestern flanks of the Wicklow Mountains, and on the opposite side of the mountains the group includes such rivers as the Luggala River above Lough Tay, the Inchavore River, the Ow River, and the rivers that flow through the six Wicklow glens lying between Glencullen in the north and Glenmalur in the south (see fig. 3.3, p.23). The rivers of the group all drain from a watershed that lies very close to the median line of the outcrop of the Leinster granite, and because of their cross-cutting relationship to the Caledonian structures, these rivers have often been regarded as consequent streams. Their origin nevertheless remains obscure, although if the entire mass of the Leinster batholith really did experience Tertiary uplift (see p.87), then the northwestward and southeastward flowing rivers might have been initiated upon the flanks of the resultant pericline. The fact that the discordant drainage today seems to be undergoing a progressive dismemberment as a result of the activities of aggressive strike streams certainly suggests that the discordant drainage is of no great antiquity.

One group of southeastward flowing rivers merits particular mention because of its puzzling nature — the group of rivers draining southeastwards from the Tullow Lowland, through the Tinahely Hills, and into the Augrim-Kildavin corridor (see fig. 3.3, p.23). The Tinahely Hills extend southwestwards from Lugnaquillia for some 30 km, and the schist and marginal granite forming the hills together rise some 280 m above the level of the Tullow Lowland. Yet along most of their length the Tinahely Hills do not form a watershed. Five rivers, including the Slaney, flow acriss the ridge in a series of narrow, steep-sided valleys, while near Moylisha [S922652] there is a dry gap which seems to have carried a similar southeastward flowing river until comparatively recently. An understanding of the drainage of this particular area must await further study, although it is difficult to avoid Farrington's (1927) conclusion that at the time when these rivers first flowed across the schist aureole, then the granite topography in the Tullow region must have stood at least as high as the summits of the present Tinahely Hills.

Rivers of the third type — the subsequent rivers, — are widely developed throughout Leinster and the following are some examples of the genre: the Dargle between the Powerscourt Waterfall and Enniskerry (flowing northeastwards in the Ordovician slates); the Vartry above Roundwood (flowing southwestwards in the Bray Group rocks, its deflection at Roundwood southeastwards into The Devil's Glen evidently being a glacial diversion); the Cloghoge River between loughs Tay and Dan (flowing southwestwards in the schist aureole); the Annamoe River between Annamoe and Laragh (flowing southwestwards in the Ordovician slates); and the River Bann in northern Co. Wexford (flowing southwestwards, again in the Ordovician slates). A particularly interesting example of subsequent drainage lies in the Aughrim — Kildavin corridor, the long strike valley extending for 32 km from Augrim to the Slaney and lying between the Tinahely Hills and the Cummer Vale Ridge (see fig. 3.3, p.23). The unexpected feature of the corridor is that it is drained in two directions — northeastwards by the Derry Water and southwestwards by the Derry River — and the heads of the two strike streams involved both lie in a marsh on the flat floor of the corridor near Tinahely. The entire valley was probably once drained southwestwards to the Slaney, and the modern presence of two subsequent rivers in the corridor is seemingly a result of glacial interference. The Derry Water and the Derry River receive as tributaries the southeastward flowing streams draining off the Tullow Lowland and through the Tinahely Hills, but in the Cummer Vale Ridge, to the southeast of the corridor, the pre-capture 'consequent' courses of the southeastward flowing rivers are perhaps marked by such prominent wind-gaps as the Wicklow Gap [T108690], the Cummer Vale Gap [T070667], and the Kilcavan Gap [T034664].

In two other areas the development of subsequent drainage has had a profound influence upon the topography: in the Liffey basin and on the Tullow Lowland. At one time the Liffey and the Kings River were two neighbouring parallel rivers draining the northwestern slopes of the Wicklow Mountains, the Liffey reaching the Central Lowland via the Brittas col [0031221] (see fig. 3.3, p.23). Later the development of strike streams in the Ordovician slates disrupted the original drainage pattern; the Liffey was diverted southwestwards into the Kings River and the Brittas col was abandoned (Farrington 1929). Today the broad floor of the resultant strike lowland is largely occupied by the Blessington Reservoir.

Farther south, the Tullow Lowland is essentially another, but even more extensive strike lowland, developed in this case along the axis of the Leinster batholith. The lowland is drained towards the southwest by the Derreen River, the left-bank tributaries of which seem to have encroached successfully upon the catchments of the rivers mentioned earlier as flowing southeastwards off the lowland and through the Tinahely Hills. Presumably, in the course of time, the Ow River will be the next major victim of the Derreen River's piratical tendencies.

Rivers in the uplands of Connacht

In Connacht, as elsewhere in Ireland, there is to be seen the phenomenon of rivers flowing off the interior lowlands and through the upland rim in order to attain the sea. The Ballysadare River, for example, between Collooney and Ballysadare in Co. Sligo, passes across the northeastern portion of the Ox Mountain inlier via a constricted valley some 2 km in length (see fig. 3.9, p.57). Farther to the southwest the River Moy flows across the same inlier about Foxford, and, most spectacular of all, the Owenmore River flows through the quartzite mountains of northwestern Mayo between Bellacorick and Bangor (see fig. 3.8, p.53). These three rivers have all been regarded as descended from ancient consequent rivers initiated upon some former, seaward-sloping sedimentary cover, but again such an interpretation must be viewed with scepticism. The discordant reach of the Ballysadare River is only short and is probably fault-guided (see p.55); the case of the Moy awaits detailed examination; and the behaviour of the Owenmore is perhaps to be attributed to the activities of westward-draining glacial meltwaters (see p.54).

Farther to the south in Connacht, Dewey and McKerrow (1963) have postulated the late-Tertiary existence of a series of eastward and northeastward flowing consequent rivers draining the flanks of the dome which they believe to have been centred upon the Maumtrasna massif (see p.87). During the Pleistocene they believe ice and meltwater to have effected major modification of the drainage pattern including the wholesale reversal of such rivers as the Erriff (McManus 1967). But in so heavily glaciated a region as the Killary Mountains all such attempts to reconstruct ancient consequent drainage lines are hazardous, and all that can be said with confidence is that today many of the valleys amidst the mountains of Galway and Mayo follow faults and other lines of structural weakness.

The rivers of Ulster

The drainage pattern of Ulster is conveniently considered under two heads: firstly the rivers of Donegal, and secondly the rivers of the Lough Neagh depression.

In Donegal many of the rivers follow the region's Caledonian grain, but there are a number of discordant rivers which flow obliquely across the underlying structures towards either the west or the east (see fig. 3.11, p.61). Examples of such westward-flowing rivers are the Clady near Bunbeg, the Owenator to the southeast of The Rosses, and the Owenea at Glenties, while the eastward-flowing rivers include among their number streams such as the lower Leannan near Rathmelton, the Swilly about Letterkenny, and the Finn near Ballybofey. Dury (1959) interpreted all these discordant rivers as the superimposed descendants of consequent streams that were initiated upon a chalk surface following early Tertiary diastrophism, one set of consequents draining towards the Atlantic and the other towards the Lough Foyle depression. He traced the former watershed between these two sets of rivers as a sinuous line extending southwards from over the site today occupied by Muckish Mountain [C004286]. Dury believes that this pattern of consequent rivers was partially obliterated in preglacial times as a result of the development of subsequent rivers along the Caledonian lines, while he sees the preglacial watershed as glacially breached in over twenty places, being locally displaced by more than 3 km and lowered by as much as 240 m (Dury 1957, 1958, 1958).

Farther to the east, in the area centred upon Lough Neagh, the drainage pattern has probably passed through three stages of development. Firstly, during immediate post-basaltic times there developed upon the basalts a centripetal drainage pattern, the drainage being focused upon the Lough Neagh depression which was then being formed by downwarping and faulting. It was these rivers which imported the debris that became the Lough Neagh clays. Secondly, this centripetal system was superimposed onto the underlying Mesozoic and older rocks as the basalts were peeled away from the country to the west and southwest of the depression. Finally, widespread dismemberment of the centripetal drainage occurred as rivers became adapted to the structures now revealed in the sub-basaltic rocks, and as vigorous obsequent rivers around the fault-guided margins of the basaltic plateau extended themselves headward at the expense of the centripetal system. Considerable adaptation to structure has occurred, for example, in the basin of the Foyle to the west of the basaltic country, while the rivers of the Antrim glens are presumably examples of streams which have encroached upon the centripetal drainage. The most striking example of encroachment, however, is afforded by the Lagan which has disrupted the centripetal drainage to the southeast of Lough Neagh by developing its own course along a northeastward striking outcrop of the Triassic sandstone. But despite such adaptation to structure, and dismember-ment, elements of the old centripetal drainage system still linger in the landscape. Lough Neagh continues to receive the waters of seven major rivers arriving from all points of the compass, while farther afield, in Co. Down, the following rivers may well be superimposed survivals of the ancient centripetal system: the Ravernet south of Lisburn, the Lagan to the southeast of Lurgan, and the Upper Bann around Banbridge. Problems nevertheless remain. The origin of the deep valley of the Newry River between the Mourne Mountains and the Slieve Gullion-Carlingford Mountains is inadequately under-stood. Glacial meltwaters seemingly gave the valley its present form (see p.69) but how did the valley originate? Equally, but farther to the north, there is the puzzling case of the rivers Main and Lower Bann (see fig. 3.12, p.67). The Main flows south into Lough Neagh and is seemingly a survival of the ancient centripetal drainage, but only a few kilometres away to the west, across Long Mountain, there lies the course of the Lower Bann, parallel to that of the Main, but flowing in the opposite direction. Not entirely convincing is the suggestion that the Lower Bann was originally a southward-flowing member of the ceripetal system, but that it was gradually reversed as a result of some structurally-guided stream encroaching steadily from the north.

One thing is certain; in eastern Ulster, as elsewhere in Ireland, glacial modification of the pre-existing drainage has been wide-spread. One of the most familiar cases of this phenomenon is the diversion of the Bush in northern Co. Antrim. The river once flowed northwards, past the layered basalt-upon-chalk structure of Knocklayd, to reach the sea at Ballycastle. In postglacial times, however, the Bush has been deflected from its original course by the Armoy Moraine (see fig. 6.2, p.121) and below Armoy the river wanders off to the west eventually entering the sea some 20 km to the west of Ballycastle (Charlesworth 1939).

Rivers in the Central Lowland

Over the greater part of the Central Lowland there has again been a good deal of glacial interference with the former drainage lines. Thick spreads of drift (at The Curragh, in Co. Kildare, drift thicknesses in excess of 70 m are known) obscure the preglacial topography and much of the present drainage pattern must be essentially postglacial. One case in point is the Liffey. In preglacial times what is now the upper Liffey left the Wicklow Mountains near Ballymore Eustace and flowed thence westwards via Kildare and Monasterevin to join the Barrow (Farrington 1929). At the conclusion of the last glacial episode the river found it impossible to resume its former course; the South Ireland End Moraine (see fig. 6.2, p.121) interposed an insurmountable barrier for the river at The Curragh. The river therefore swung northwards at Kilcullen, and then north-eastwards, to reach the sea at Dublin Bay, thus bringing the modern Liffey into existence.

A large part of the Central Lowland is of course drained by the Shannon, a river which, as we have seen, many authors have regarded as a superimposed descendant of an ancient southward-flowing consequent river. Again, however, such a simplistic interpretation must be treated with reserve. It seems more likely that the Shannon is really rather young as a river, the result of the postglacial linkage of a series of glacially excavated rock-basins such as loughs Ree and Derg. But this explanation leaves unresolved the problem of the Shannon's gorge at Killaloe (see fig. 3.2, p.19). No feature of the Shannon's course has done more to sustain the superimposition hypothesis, but such studies of the Shannon as have been made suggest that the history of the gorge has been rather more complex than the simple superimposition hypothesis would allow (Kilroe 1907b; Farrington 1968). It now seems that in preglacial times the proto-Shannon flowed, not through the Killaloe Gorge, but through the limestone-floored corridor around Scarriff to the west of Lough Derg and between the Slieve Aughty and Slieve Bernagh inliers. At that time there was a low col at the southern end of Lough Derg at

Rinnaman Point, but this col has now been breached by some process (glacial erosion is the obvious candidate), thus allowing the river to abandon the Scarriff corridor in favour of its spectacular exit from Lough Derg via the Killaloe Gorge. A preglacial river certainly had its head in the vicinity of Killaloe, but the modern Shannon seems to have missed the buried channel of its preglacial ancestor and between Castleconnell and Limerick there are a number of rapids in the modern river where it encounters the rock-walls of the ancient valley (Lamplugh *et al*. 1907, pp.2-3). The course of the Shannon, nevertheless, still contains many puzzles; it remains a subject that would repay further study.

6 Pleistocene events

It was in 1867 that Close published the earliest essay in Ireland's glacial history, and since that date a considerable literature has developed as successive generations of geologists have sought to unravel that tangled skein which is the Irish Pleistocene. By the 1950s the work of Charlesworth and Farrington had established that ice-sheets had covered the greater part of the island on at least two occasions; the older of these glacial periods Farrington termed 'the Eastern General', while the younger he styled 'the Midland General'. In the last few years new terms have appeared (table 6.1 and fig. 6.1), 'Munsterian' replacing 'Eastern General', and 'Midlandian' replacing 'Midland General'. Several interglacial deposits of Gortian age (Jessen *et al.* 1959; Watts 1970) lie below the Munsterian glacial deposits and are regarded as broadly equivalent to the British Hoxnian. Only one Last Interglacial (Ipswichian) deposit is known, that being at Shortalstown, Co. Wexford, where it separates the underlying Munsterian deposits from the overlying Midlandian. Elsewhere the Munsterian and Midlandian glaciations have been identified by the litho-stratigraphic succession of deposits, and distinguished by techniques such as till-fabric analysis, erratic content, degree of weathering, and, to some extent, by differences in the surface morphology of the various drifts.

Before embarking upon a description of the distribution, chronological sequence and morphological expression of the Munsterian and Midlandian drifts, it must be emphasised that our knowledge of these deposits, and of the glaciations responsible for them, is still far from satisfactory. The same applies in even greater measure to certain older sediments found below the Munsterian drifts. Correlations of glacial events at different periods across the country are still tentative, and only generalised relationships of ice movement can be shown as in fig. 6.1, this being especially true for the Munsterian. For the latter part of the Midlandian, however, pollen analysis and 14C dating techniques come to our aid making possible a far more detailed understanding of the many stages of glaciation and deglaciation.

The Munsterian glaciations

The earliest known general glaciation involved the expansion of ice from a number of centres (fig. 6.1). A powerful ice-sheet of Scottish origin moved across northeastern Ulster in a southwesterly direction, penetrating as far as the Lough Neagh basin and Lough Foyle, and perhaps skirting the northwestern coast of Co. Donegal to the tip of the Slieve League peninsula at Malin Beg. This ice deposited a chocolate-coloured or brownish-grey till, generally containing few stones, but displaying many fragments of marine shells. The extent of this early Scottish, or Irish Sea ice, was restricted in places by the peripheral uplands as well as by the presence of large ice-masses which had built up within Ireland. However, shelly calcareous till is seen along the east coast from Co. Down to Co. Wexford, and as far west as Cork Harbour. Mitchell (1962) referred to the deposit as the Ballycroneen till, from the type site in Co. Cork, but it has since been realised that several facies of shelly till are present on the eastern and southeastern coastlands (Mitchell 1972; Stephens *et al.* 1975).

The Munsterian General ice appears to have built up as a great internal ice-sheet, with an axis running from the Lough Neagh basin, southwestwards to Connemara. Ice movement was apparently powerful enough to carry granite from Connemara southeastwards across Munster. It was also able to bring this inland ice into contact with Irish Sea ice near Drogheda (Colhoun and McCabe 1973; McCabe 1972), and to the south of the present coastline between

---

Fig. 6.1 Interglacial sites (A) and general ice movements during the Munsterian (B) and Midlandian (C and D) glacial periods.

(B) (1) Southwestern limit of Irish inland ice (Munster General glaciation).
     (2) Western limit of coastal Irish Sea ice.
     (3) Outer limit of the Greater Cork-Kerry glaciation.
(C) and (D) Midlandian — there are still difficulties in assigning limits in Inishowen, on the west coast and near Cork city.

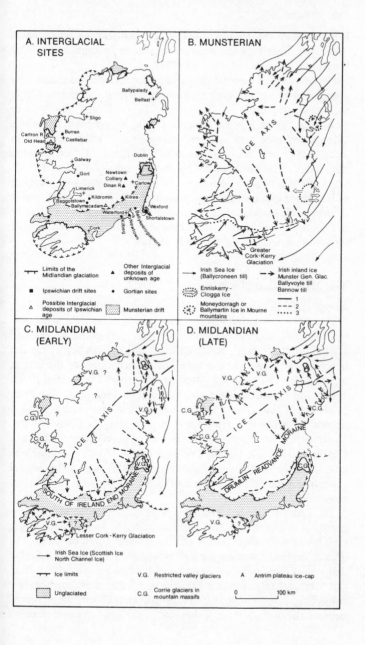

A. INTERGLACIAL SITES

Ballypalady ▲
Belfast +
+ Sligo
Cartron R △
Old Head △
■ Burren
+ Castlebar
+ Galway
Dublin
+ Gort
Newtown Colliery ▲ Dinan R ▲
+ Carlow
+ Limerick
Kildromin ●
Baggotstown ■
Ballymacadam ▲ ● Kilree
Wexford
Waterford + Ballylinegormore
Newtown Shortalstown
+ Killing
+ Cork

Limits of the Midlandian glaciation
▲ Other Interglacial deposits of unknown age
■ Ipswichian drift sites
● Gortian sites
△ Possible Interglacial deposits of Ipswichian age
Munsterian drift

B. MUNSTERIAN

ICE AXIS

Greater Cork-Kerry Glaciation

→ Irish Sea Ice (Ballycroneen till)
⟶ Irish inland ice Munster Gen. Glac. Ballyvoyle till Bannow till
⋯⋯ Enniskerry - Clogga Ice
⣿ Moneydorragh or Ballymartin Ice in Mourne mountains
— 1
– – 2
⋯⋯ 3

C. MIDLANDIAN (EARLY)

V.G. ?
V.G.
C.G.
C.G.
ICE AXIS
V.G.
SOUTH OF IRELAND END MORAINE
V.G.
Lesser Cork - Kerry Glaciation

D. MIDLANDIAN (LATE)

V.G.
C.G.
ICE AXIS
V.G.
DRUMLIN READVANCE MORAINE
C.G.
V.G.

→ Irish Sea Ice (Scottish Ice North Channel Ice)
⊤⊤ Ice limits
Unglaciated
V.G. Restricted valley glaciers
C.G. Corrie glaciers in mountain massifs
A Antrim plateau ice-cap

0   100 km

Kilmore Quay and Dungarvan. It is also possible that there was a zone of contact between the two ice-masses in southeastern Co. Wexford (Colhoun and Mitchell 1971).

The Munsterian General ice perhaps accumulated to a considerable thickness by the amalgamation of different ice-centres within the island, especially in the Central Lowland and the western mountains. There is some evidence for the existence of important independent mountain ice-caps, or systems of valley glaciers, in areas such as the Wicklow Mountains (Synge 1973), the Mourne Mountains (Stephens *et al.* 1975), and the mountains of west Cork and Kerry (Farrington 1954; Lewis 1974). In the northwest, deposits of weathered and heavily cryoturbated 'local' till are known, and it is conceivable — indeed likely — that there was competition in counties Donegal, Londonderry and Tyrone between the inland ice and the invading Scottish ice (Colhoun 1971a).

The Midlandian glaciations

At this stage the pattern of glaciation from a number of different centres of accumulation was repeated, although the ice-masses were less extensive and probably thinner than during the Munsterian glaciations. Scottish ice, North Channel ice and Irish Sea ice did not penetrate far across the northern and eastern coasts of the island, reaching only to the Armoy Moraine in northern Co. Antrim (Prior 1968), and to northeastern Co. Down (Hill 1968; Hill and Prior 1968), while farther south the ice may have been restricted to a narrow coastal fringe a few kilometres wide along the Leinster coast. Contrary to the 'traditional' views of Dwerryhouse (1923) and Charlesworth (1939, 1963b), it now seems that this ice was unable to cross the Antrim Plateau. Equally it probably was not powerful enough to penetrate to the west of either the Mourne or Wicklow mountains.

The limits of this last glaciation have been mapped using stratigraphical and morphological evidence, aided by relationships which exist between the parent drifts and the soils produced on them. The greater degree of weathering and cryoturbation of the Munsterian drifts have also been used in attempts to separate the 'older' from the 'younger' tills. While it can be argued that these methods are subjective, considerable differences can undoubtedly be observed between the drift-sheets on either side of the great terminal moraine

Table 6.1   Late Pleistocene events in Leinster

General names	Farrington (1944, 1954, 1966a)	Mitchell (1973)	Synge (1973) Midland Ice	Mountain Ice
Last glaciation or Midlandian glaciation (British Devensian)	Mountain glaciation; Midland General glaciation; Athdown Mt. glaciation	Killiney and Shortalstown upper shelly tills; Athdown Moraine	Colbinstown Blessington Hacketstown Irish Sea (Glenealy) till on E. Coast	Late Athdown Early Athdown Brittas Aughrim
Last Interglacial (British Ipswichian)	Weathering horizons developed on surfaces of older drifts	Shortalstown estuarine sand	Clogga gravel represents a raised-beach resting on a planated surface of lower till, a surface laterally continuous with a rock platform	
Munsterian glaciation (British Wolstonian)	Brittas Mt. glaciation; Eastern General glaciation, (Irish Sea Ice); Enniskerry/ Clogga Mt. glaciation; Solifluction	Brittas Mt. glaciation; Ballyvoyle till; Bannow till; Killiney and Shortalstown lower shelly tills; Enniskerry granite till; Head at Kilmore Quay	Clogga till Ballyvoyle till } Bannow till Ballycroneen shelly till (Irish Sea Ice) Mt. Glaciation?	from the Irish Midlands
Gortian Interglacial (British Hoxnian)	Gort Interglacial; Raised-beach gravels at some points on Wexford, Waterford and Cork coasts ? ?	Ballykeeroguemore mud; Kilmore Quay and Wood Village raised-beaches	Raised-beach gravels? Wave-cut rock platforms?	
Pre-Gortian	Wave-cut rock platforms	Erratics reach south coast of Ireland; Wave-cut rock platforms	Wave-cut rock platforms?	

1 In the left-hand column the British equivalents of the general Irish subdivisions are given, without implying exact correlation in time.

2 No exact correlation is implied between deposits or glacial events which happen to be printed level with one another in different columns.

3 Stratigraphical order is attempted in each 'box', where known, but some successions are still uncertain, especially during the Munsterian glaciation.

Source: Based on published works of Farrington, Mitchell and Synge.

which swings round the northern end of the Wicklow Mountains and across Ireland through Tipperary to the mouth of the Shannon. It is known as the Southern Ireland End Moraine or Tipperary Line, but it should more properly be called the Ballylanders Moraine from the type site in Co. Limerick. To the north of this line there is fresh, relatively unweathered till at the surface, lacking any sign of the conspicuous cryoturbation features (e.g., the wholesale erection of stones to depths of 2 to 3 m) which are common where the 'older' drifts are exposed at the surface. Furthermore, to the north of the line there are well-marked topographical features, such as kames, moraines, eskers, and drumlins, which are infrequently seen in the 'older drift' country.

Although three-quarters of the island was covered at this stage, the ice was not powerful enough to engulf all the peripheral mountains, and in the southwest ice from the Irish Midlands failed to make contact with the Lesser Cork-Kerry ice. Figure 6.2 is based upon a map prepared by Synge (1970) in which four major stages of the Midlandian glaciations are depicted:

Stage A   The Southern Ireland, Tipperary, or Ballylanders End-Moraine.
Stage B   The Galtrim Moraine.
Stage C   The Drumlin (Kells, Fedamore, Killough, Armoy, Moville) re-advance moraines.
Stage D   The Kilrea and Lisburn-Dunmurry moraines.

Consideration of figs. 6.1 and 6.2 shows that the maximum limits of the glaciation were not everywhere contemporaneous, although as yet the absolute dating of each stage has not been achieved. At Stage B, Synge (1950) has shown that the northeastern part of the Galtrim Moraine represented a still-stand phase, whereas to the southwest of the Trim area, where the moraine breaks the continuity of the esker-chains, there is some evidence for a slight oscillation or re-advance of the ice. Synge (1950) has shown that, provided the annual character of the esker 'beads' is accepted, then the ice retreated from the Galtrim Moraine at a rate varying from 130 m per year (Laracor esker) to 76 m per year (Ballinrig esker). In eastern Co. Down, at

---

Fig. 6.2   Quaternary geology, based largely upon the records and fieldwork of the Geological and Soil Surveys of the Republic of Ireland and the published works of Charlesworth, Colhoun, Farrington, McCabe, Mitchell, Stephens, Synge, and Watts.

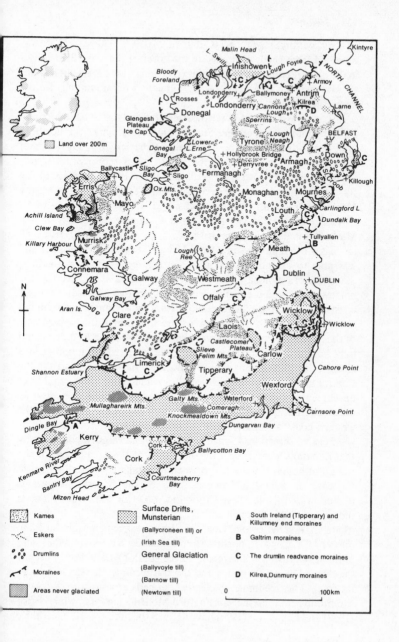

Land over 200m

Symbol	Legend
Kames	Surface Drifts, Munsterian
Eskers	(Ballycroneen till) or
Drumlins	(Irish Sea till)
Moraines	General Glaciation
Areas never glaciated	(Ballyvoyle till)
	(Bannow till)
	(Newtown till)

A South Ireland (Tipperary) and Killumney end moraines

B Galtrim moraines

C The drumlin readvance moraines

D Kilrea, Dunmurry moraines

0   100km

Glastry, within Stage C, a 14C date of 24 050 ± 650 years B.P. was obtained from shells contained in the lower of the two tills involved in the construction of the drumlins (Hill and Prior 1968), but it is not yet known if this represents a true, or merely a minimum, age for the lower till member.

It seems possible that in northeastern Ireland the so-called Armoy Moraine may also represent Stage C, in that it marks the northern limit of the drumlins in the Bann valley (Charlesworth 1939, 1963b; Prior 1968), even though at the same time it represents a south-westerly limit of Scottish ice (Hill and Prior 1968). Irish ice in the Bann valley may have been in contact with Scottish ice at the Armoy Moraine (Creighton 1974; Stephens *et al.* 1975).

Recognition of Stages A, B, and C implies that early in the Midlandian glaciation the Irish ice was moving away from an ice-axis located along a line from Lough Neagh to Lough Ree, and was able to extend furthest in a southeasterly direction (fig. 6.2, Stage A). In contrast, at Stage C the ice-axis shifted to the northwest, and consequently pressure increased towards the west coast. Perhaps this change resulted from an asymmetry developing along the ice-axis following heavier precipitation in the western part of the island. Certainly moisture-bearing air from the Atlantic would have had no difficulty in penetrating the scattered mountain groups along the western seaboard, and snow must have accumulated in the Central Lowland. Under such conditions a large ice-sheet could have developed within the discontinuous rim of mountains. Ice certainly did not build up initially in the separate mountain groups on the western coast as Charlesworth (1963b) suggested, to radiate out-wards in all directions. Synge (1968, 1970) has argued convincingly that nunataks were present in some parts of the upland rim of the island, although the survival of such nunataks is more doubtful in the case of inland hill masses such as the Sperrin Mountains (Colhoun 1970), and Slieve Croob in Co. Down. Fieldwork suggests the presence of unglaciated areas in Co. Donegal in Inishowen (Stephens and Synge 1965) and the Slieve League peninsula, in small areas of Co. Sligo, in northwestern Mayo (Synge 1963b, 1968), and in the Killary Mountains region to the south of Clew Bay. Parts of the Slieve Felim Mountains and the Castlecomer Plateau were also ice-free. The ice-limit is clearly defined around the Wicklow Mountains and in the Carlingford-Mourne massifs, while Prior (1968) and Creighton (1974) have defined the limits of the general ice in part of northern and northeastern Co. Antrim. But there has been revision of the extent of these so-called ice-free areas, some of which have been shown to have supported small local ice-caps (e.g., the Glengesh Plateau in southwestern Co. Donegal), while in others the general ice is believed to have extended farther than was formerly thought.

Interglacial deposits

The position of the known interglacial sites is shown in fig. 6.1a, but the distribution conveys nothing of the controversy concerning both the Gortian and Ipswichian deposits. There is no really satisfactory type-site for the Last or Ipswichian Interglacial, so for the present the Shortalstown estuarine sand, from Shortalstown, Co. Wexford, must serve as the sole representative.

Gortian Interglacial deposits

These deposits have been examined principally by Jessen *et al.* (1959), Watts (1959, 1964, 1967) and Mitchell (1970a, 1976). At each site a temperate mud, peat, or plant bed was sealed below glacial till, or tills, by outwash gravels, by head deposits, or by a combination of all three. Several of these sites lie to the south of the Southern Ireland End Moraine. Consequently, on a strictly stratigraphical basis, the temperate deposits are regarded as pre-dating several different facies of Older Drift. A strong case has been made (Watts 1970; Mitchell 1970a) for regarding these widely separated temperate deposits as belonging to the same general group (termed Gortian) and for there to be a reasonable correlation with the British Hoxnian. But doubts must linger. It is above all the absence of Irish sites of Ipswichian age which still casts doubt upon the Gortian-Hoxnian correlation. If there is a handful of deeply buried Gortian sites known, then why is there not at least an equal number of known Ipswichian sites? The latter would not necessarily be so deeply buried, and in any case there are many good, deep sections available for examination in drifts assigned to the Midlandian glaciation. Furthermore, if the Irish Midlands had sustained peat growth during the Last Interglacial to anything like the degree they do today, then the most recent till sheets should bury and contain copious quantities of organic material. But they do not, and, despite considerable field investigation, all that has been found are the presumed Gortian deposits. The following are the most accessible of the Gortian sites: Ballykeerogemore, Co. Wexford [S7319] (Mitchell *et al*. 1973); Newtown, Co. Waterford [X7006] (Mitchell 1970a); between Fenit and Spa, Co. Kerry [Q7714]; Boleyneendorrish, near Gort, Co. Galway [M5305].

Shortalstown Interglacial deposits

Near Shortalstown House, Co. Wexford [T0214], at about 30 m O.D., temporary land drainage trenches in glacial deposits revealed a complex series of sections where an Interglacial estuarine sand, with associated beach gravels, was disturbed and interbedded between an upper shelly till and a lower shelly till (Colhoun and Mitchell 1971). The site lies 1.5 km inside the revised southern limit of the Midlandian ice (see fig. 6.2, p.121), outside which Munsterian till is deeply disturbed by frost heaving, solifluction, and the formation of pingos (Mitchell 1971). The pollen diagram shows *Betula, Pinus* and *Quercus* dominant, together with a presence throughout of up to 5% of *Ulmus* pollen. There was little non-arboreal pollen and it was concluded that a temperate forested landscape was present. The vegetation compares closely with Zone *e* at the type site for the Last Interglacial in Britain, at Bobbitshole, Ipswich (West 1957), and although having only a short pollen diagram, the Shortalstown pollen record differs from all the known Gortian sites.

The 'pre-glacial' beach and associated Head deposits

At places on the coast of counties Cork, Kerry, Waterford and Wexford beach gravels, consisting mainly of local rock types, but with some erratics, rest upon a well-planed rock-platform, and are deeply buried below till and head deposits. Various quantities of angular and sub-angular material may be present and mixed in with well-rounded gravel and sand, but no marine shells have been recorded (see fig. 6.2, p.121). The association of the gravels with an abrasion platform and former cliff-line is well displayed just east of Howe's Strand in Courtmacsherry Bay, Co. Cork, near Garryvoe in Ballycotton Bay, Co. Cork; at Wood Village near Fethard, Co. Wexford; and at Kilmore Quay, Co. Wexford. The cliff-line has been observed to pass from rock to glacial drift at the western end of Whiting Bay, Co. Waterford. The mixture of angular rock fragments with beach gravels has been described from several sites. Although the rapid erosion of pre-existing till and head deposits, together with chance cliff falls, may account for the incorporation of angular debris, there is also the possibility that the gravels may not always be of marine origin, or that severe frost shattering was operating on exposed rock surfaces while beach construction was taking place (Bryant 1966; Farrington 1966b; Stephens 1970; Wright and Muff 1904). However, most authorities would agree that at many sites on the southern coast of Ireland, and at a few places on the Wicklow and Wexford coasts, a wave-planated rock-platform is overlain by beach gravels, which are deeply buried below till and head deposits. It seems likely that the age of the beach may vary from place to place, but it is generally regarded as being no younger than the Last Interglacial. Along the east coast the rock-platform is frequently striated, and at Wicklow Head and Arklow Head, Synge has established the presence of an elevated notch in rock at 14 m O.D. On the south coast at Newtown, Co. Waterford, part of a platform is seen to be severely shattered and the rock fragments have been churned by cryoturbation. A wave-planated rock-surface is also known on the southern side of the Howth peninsula, between Sutton and Bottle Quay, where it is overlain by two facies of till. But a postglacial beach-deposit also rests directly upon the much older striated rock-surface (Stephens and Synge 1958). Such rock-platforms are intermittently revealed below glacial deposits as far north as Belfast Lough, where the platform and cliff notch is partially buried below the drumlin-forming tills at Groomsport, Co. Down (Stephens 1957). At no sites north of Clogga have the older sub-drift beach deposits been located in contact with the platforms and this is perhaps a measure of the scouring of the rock-surface by ice, subsequent to marine planation and beach deposition. It seems likely that the altitude of the rock-platforms and their inner cliff-notches varies according to position, to exposure, and perhaps to rock type.

Pleistocene geomorphology in the four provinces

The more important aspects of the Pleistocene geomorphology of the four provinces are next considered, taking first Leinster where

Farrington laid the foundations for an understanding of the Irish Quaternary record.

Leinster (*apart from counties Louth and Meath*)

Recent investigations by Synge and his colleagues of the Geological Survey of Ireland have established a new chronology for Pleistocene events in Leinster (figs. 6.3, 6.4, 6.5, 6.6 and 6.7). Analysis both of patterns of striae and of the erratic content of various drifts show that the Clogga till is the oldest unit and that it was formed by ice moving southeastwards across the Derry Corridor of southern Co. Wicklow to reach and cross the present coastline. Farrington (1954) equated the Clogga advance with his Enniskerry Mountain glaciation, but it now seems likely that an ice-sheet moving eastwards from the Irish Midlands was also involved. The Clogga drifts are well exposed only along the east coast between Wicklow and Cahore Point, where they may consist of granite-rich till or poorly sorted gravels of mainly local rocks. No fresh depositional drift forms are developed in Clogga drift, and its surface generally carries a distinct weathering horizon. For instance, at Seabank Point many of the erratics are completely weathered, although perfectly fresh examples of the same rock-types occur lower down (Synge 1964). Desiccation-cracks or frost-cracks penetrate downwards from the surface of the Clogga till, and similar periglacial and weathering phenomena are known at many sites where these older tills are exposed. The gravel layer which is sometimes seen to separate the lower Clogga till from the overlying Irish Sea till (e.g., at Clogga Point) appears in places to rest upon a planated surface which may be a lateral continuation with the basal rock abrasion platform. The gravel is composed of stones derived from bedrock and the Clogga till, and Synge has tentatively accorded it the status of an interglacial beach deposit. Thus the gravel layer, and the abraded rock-platform on which it can be seen to rest, are envisaged as belonging to the Last (Ipswichian) Interglacial (fig. 6.4). This interpretation places the ice-moulded and striated rock-platform of the east coast of Ireland between Wexford and Belfast Lough (Stephens 1957) in a stratigraphical position such that it does *not* pre-date all the known Irish glacial drifts.

In Co. Wicklow to the east of the mountains, and in eastern Co. Wexford, there are in places two shelly till members. As we have already seen there is at Shortalstown, Co. Wexford, an Interglacial

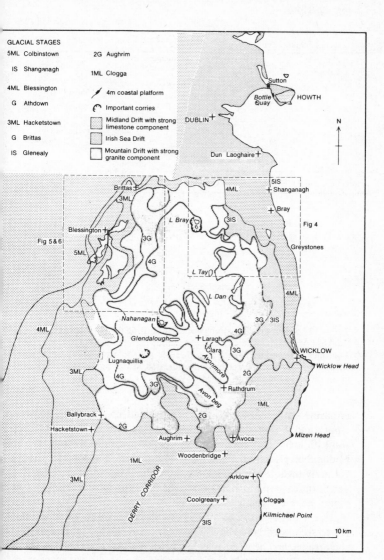

Fig. 6.3 Glacial stages and drift types in the Wicklow Mountains and adjacent areas, based upon the work of Farrington and Synge, and redrawn from a map by Synge.

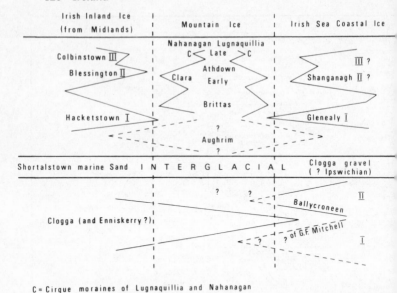

Irish Inland Ice (from Midlands)	Mountain Ice	Irish Sea Coastal Ice

C = Cirque moraines of Lugnaquillia and Nahanagan

Table 6.2   A tentative chronology of events during the late Pleistocene in the Wicklow Mountains and adjacent coastlands between Dublin and Wexford. The question marks indicate that there is still an element of uncertainty concerning some of the correlations indicated in the figure, which is based upon the latest findings of Synge.

estuarine sand, considered to be Ipswichian in age, lying between Upper and Lower shelly tills (Colhoun and Mitchell 1971). Thus some till members containing fresh limestone probably indicate the Midlandian glaciation and not the Munsterian, and some of the till formerly regarded as originating from the invasion of Irish Sea ice across the eastern coastline during the penultimate glaciation may in reality be much younger.

Recent work (Synge 1973) has established a distinct inland limit for fresh, unweathered limestone drift (with chert, flint and marine shell fragments) between Wicklow town and Arklow. Farther south the Older Drift extends southwestwards as a broad belt of country including the Derry Corridor and the surrounding hill areas lying between the outer limits of the Midlandian inland ice pressing southeastwards, and the Midlandian Irish Sea or coastal ice covering the low coastal plain to approximately the 90-m contour between Wicklow and Gorey. The older Irish Sea, or Munsterian shelly calcareous till, is known near Sutton, Co. Dublin (Stephens and Synge 1958), at Killiney, Co. Dublin, and as a thick deposit along the east coast south of Wicklow Head. Frequently it is deeply weathered, the natural grey-brown colour oxidised to red-brown with complete decalcification to depths varying from 2 to 4 m.

The approximate western limits of the ice-sheet are shown in fig. 6.2. There is some evidence that the Wicklow Mountains were not entirely engulfed, although the ice reached 550 m at the northeastern end of the massif. Farrington (1949) recognised the till 24 km inland from the coast on Saggart Hill at 305 m, and from the records of marine shells at sites in the Central Lowland it is possible that the Irish Sea ice extended well inland from the coast. The ice-margin declined from about 152 m near Wicklow town to 61 m near Arklow. Farrington also suggested that because of the presence of a small percentage of Irish Sea erratics, the early Enniskerry Mountain ice-cap and hence the Clogga (general) ice-sheet, were at one stage in contact with the Irish Sea ice in eastern Co. Wicklow. Local mountain ice may have prevented the Irish Sea ice from extending higher than 550 m. But because the Irish Sea (Ballycroneen) till always overlies the Clogga till, this must indicate that the former was the more powerful.

The waning of the Munsterian ice released enormous amounts of meltwater, which possibly accounted for a number of water-cut gorges of enormous size, as for example, The Scalp and The Glen of the Downs in northeastern Co. Wicklow.

We turn now to events attributable to the Midlandian glaciation. The Aughrim advance (fig. 6.4) is associated with scatters of huge perched blocks of granite (e.g., on Trooperstown Hill, on the northern flank of Cushbawn, and at Ballybrack), the limits of which were established by Farrington (1954) and subsequently revised by Synge (1973). Synge has demonstrated that the Aughrim advance was restricted to the expansion of local mountain glaciers, such as the Avonbeg glacier with its strongly marked morainic limit at Avoca. At Aughrim deltas at 85 and 88 m mark the site of ponded water-bodies, fed by outwash from the Ow valley glacier and a western lobe of the Avonbeg glacier.

Local mountain ice at the Aughrim stage does not appear to have been in direct contact with any ice-sheet encircling the Wicklow

Mountains. Synge (1973) considers that the outer limit of the local ice pre-dated both the Midland ice advance to Hacketstown, and the Irish Sea coastal ice pressing southwards against the mountain flank in Glenealy, west of Wicklow town. It should be noted, however, that as indicated in fig. 6.4, the Aughrim advance ice extended into the area subsequently invaded by Irish Sea coastal ice. Thus, there is at present little unequivocal evidence to allow the extensive Aughrim mountain ice advance to be correctly placed in the chronology. Synge (1973) favours an early Midlandian age, but the matter appears open and it is not impossible for the Brittas and Athdown advances to mark the only significant expansions of mountain ice during the Last (Midlandian) Glaciation.

The maximum advance of the Irish Sea coastal ice invading Glenealy, sweeping across Wicklow Head and around Ballymoyle Hill to Shelton Abbey and beyond, is associated with severe peri-glacial conditions. Thick head deposits mantle some hill-slopes outside the limits of the Irish Sea and local mountain ice-masses, whereas in the Three Mile Water valley, in the Redcross basin, and in the Snugborough basin, pingos are found close to, but outside the Irish Sea drift-limit. Mitchell (1971) has reported a similar relation-ship in southern Co. Wexford. It is interesting to observe (fig. 6.2; Synge 1973) that pingos also occur at Cronybyrne and Glasnarget inside the plotted limits of the Aughrim advance, but outside both the Glenealy (Irish Sea ice-maximum) limit and the Ballinaclash stage of the Avonmore glacier, which extended to Rathdrum. It is not yet clear if this evidence has any real bearing upon the age of the Aughrim advance.

The Wicklow Mountains   In the area delimited by fig. 6.4 the relationship of the various Midlandian glacial-limits of both local mountain and inland ice can be observed. Farrington (1944) carried out some of his classic work around Enniskerry, where he identified a series of different till and gravel deposits involving the two general advances of the Munsterian and Midlandian ice, together with the related stages of mountain glacier expansion (see table 6.2, p.128). The area has been re-examined by Synge and his colleagues, and the map (fig. 6.4) seeks to show the position of the various ice-fronts, as well as the sequence of retreat phenomena.

The upper limit of Midlandian General ice is indicated by the presence of fresh limestone till to line 3IS (Glenealy stage), which is

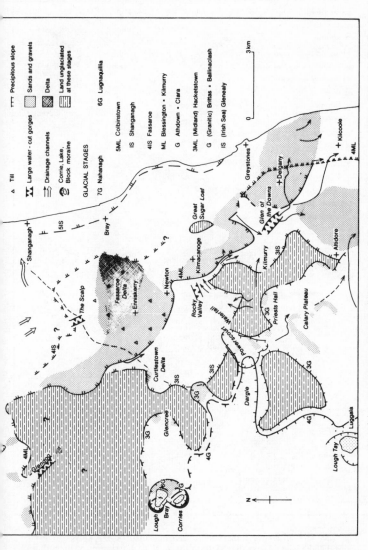

Fig. 6.4   Glacial stages in the north eastern part of the Wicklow Mountains, based upon the work of Farrington and Synge, and redrawn from a map by Synge.

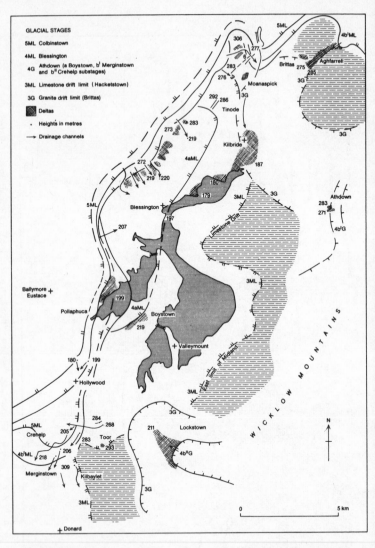

Fig. 6.5 Glacial stages in the northwestern part of the Wicklow Mountains around Lake Blessington, based upon the work of Farrington, Mitchell, and Synge, and redrawn from a map by Synge.

regarded as equivalent to line 3ML (Hacketstown stage) shown in figs. 6.3 and 6.5. This general ice encircled the northeastern flank of the Wicklow Mountains while various valley glaciers in Glencree and the Dargle valley expanded to bring granite debris to lower levels. It should be noted that even during the maximum extent of the ice-masses at these various stages, considerable areas of the mountains were free of ice, and the Great Sugar Loaf stood as a nunatak above the encircling ice. The approximate relationship of the various stages to one another is shown in table 6.2, p.128.

The Fassaroe delta is a large deposit in the valley of the Dargle near Enniskerry, the gravels containing a high percentage (32-55%) of limestone and only a low percentage (less than 5%) of granite. It is not yet known if the water body which controlled the height and position of the delta-terrace was impounded by Irish Sea ice at the 4IS stage, or if it represents an earlier stage, subsequently exposed during recession. On the other hand the Curtlestown delta probably resulted from Irish Sea ice (3IS stage) impounding a lake in part of lower Glencree.

A fine arcuate moraine with outwash extends beyond the foot of Powerscourt waterfall, and represents deposition from a local mountain glacier in the Dargle valley at the 4G (Athdown) stage. The position of a number of large meltwater channels is also shown, some of which were probably used on more than one occasion, although The Scalp may not have been used during the Midlandian glaciation.

At the head of Glencree a fine suite of moraines loops across the lip of the two Lough Bray cirques. The massive outer moraine (stage 6G), about 1 km wide, is regarded as the late-glacial Athdown stage, while the fresh, double morainic ridges inside the older moraine probably represent the Zone III or Nahanagan stage (Warren 1970).

Impressive sequences of glacial deposits are revealed in the coastal cliffs north of Greystones, but especially at Shanganagh, where 4 km of sections are visible. Somewhat different interpretations of this fine and important section have been made (Hoare 1975), but there is no doubt that advances and withdrawals of inland, as well as Irish Sea ice-masses, have accounted for the sequence of tills and gravel horizons.

In the area delimited by fig. 6.5 (p.132) different phases of ice advance and retreat of the Midlandian glaciation have been revealed by Farrington (1957b), Farrington and Mitchell (1973), and by Synge and his colleagues. The topographical situation on the northwestern flank of the Wicklow Mountains was such as to allow mountain glaciers to advance along pre-existing river valleys (e.g. the Kings River), and downslope towards Lake Blessington. These glaciers carried granite drift to Moanaspick and near Lockstown, where mounds mark the Brittas limit (3G). The important Athdown limit (4G) is marked by a distinct moraine and delta (211 m) at Lockstown, and at the type site — Athdown — by a terminal moraine and the highest delta recorded (283 m) for Glacial Lake Blessington (fig. 6.6). The fluctuations in the levels of Glacial Lake Blessington depended upon the position of the general Midlandian ice-fronts (3, 4, 5ML), which may have involved ice pressing south and southeastwards from the Irish Midlands, and perhaps also southwestwards from the Irish Sea.

Three stages in the evolution of Glacial Lake Blessington are shown in fig. 6.6 which should be compared with fig. 6.5, the sequence having been worked out by studying the height relationships of moraines, deltas and various meltwater channels. As the inland ice withdrew, the area of the lake expanded to its maximum extent at Blessington stage 4ML, which is approximately equivalent in time to Athdown mountain stage 4G. Later, as the mountain glaciers disappeared from the vicinity of the lake, and the ice-sheet wasted and withdrew further westwards, the lake began to assume approximately the dimensions of the present reservoir — the Colbinstown stage 5ML. Various fluctuations in the position of the ice-fronts (3-5ML), involving withdrawals and re-advance, have been worked out in detail by Synge.

The last glacial episode in the Wicklow Mountains has been recorded at Lough Nahanagan (see fig. 6.3, p.127). The site has been studied by Colhoun. The outer massive cirque moraine enclosing Lough Nahanagan is regarded as representing the maximum extent of the cirque glacier at the head of Glendasan during the Athdown glacial stage, and it has been correlated with similar enclosing moraines at the Lough Bray cirques (stage 6G in fig. 6.4, p.131). However, three series of inner cirque morainic ridges, consisting of deformed lake-clays and granite debris, form fresh morphological features, indicating an active though short-lived cirque glaciation within the original basin. 14C dating of the ice-deformed lake-clays has yielded dates of 11 600 and 11 500 years B.P. This suggests that a cirque glacier redeveloped in Lough Nahanagan sometime after

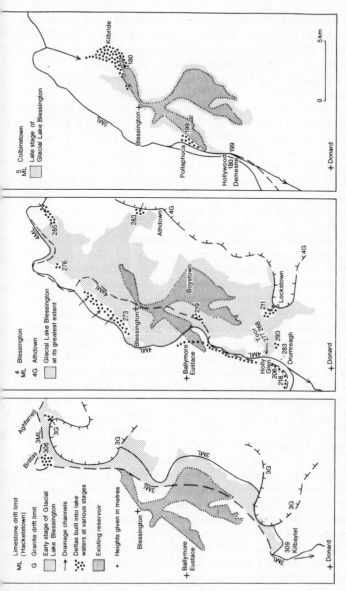

Fig. 6.6 Glacial Lake Blessington — three stages in the development of a proglacial lake, based upon the work of Farrington, Mitchell, and Synge, and redrawn from a map by Synge.

11 500 years B.P., and can probably be attributed to Pollen Zone III time, for lake-muds resting upon the fresh ridges provide a complete postglacial pollen record. One can perhaps infer from this evidence that the inner morainic ridges (stage 7G in fig. 6.4, p.131) at the Lough Bray corries, the possible late cirque glaciers of Lugnaquillia (Farrington 1966a), on Achill Island (Farrington 1953b), in the Nephin Beg range (Synge 1963b, 1969), in the Galty Mountains (Synge 1970), in the mountains of western Co. Kerry (Lewis 1974), in Co. Donegal, and in the Mourne Mountains of Co. Down, may all date from Zone III. Pollen Zone III is generally taken to extend from about 11 000 to 10 250 years B.P.

Glendalough is the next valley to the south of Lough Nahanagan and Glendasan, and it is well known as a deeply incised glacial trough (see fig. 6.3, p.127). The monastic settlement stands on a delta associated with drainage down Glendasan into a temporary lake impounded by dead ice and moraine deposits at Laragh. Two lakes now occupy the over-deepened rock-floor of Glendalough, an alluvial fan having divided the original lake into two portions.

West of the Wicklow Mountains the Blessington Moraine (4ML, fig. 6.3, p.127) extends southwards and becomes less distinctive morphologically. In the south this moraine — the Tipperary or Ballylanders Moraine — everywhere marks the outer limit of a sheet of only slightly weathered calcareous till, and it is indicated as stage A in fig. 6.2. Large outwash terraces can be seen in the valleys of the Barrow and Slaney, leading away from the ice-front position at the time of the maximum extent of the Midlandian glaciation.

In the northern Wicklow Mountains the terminal moraine of the Midlandian ice partially blocks the mouths of some of the north-trending valleys, and at Killakee, Co. Dublin, the maximum altitudinal limit of the moraine is 380 m. In Glenasmole granite outwash of Brittas age cuts through the terminal moraine of the general ice which pressed southwards into the lower part of the valley.

The limiting moraine of the general ice curls around the northern end of the mountains as line 3IS on fig. 6.3 (p.127), and the outer limit of drift with fresh limestone erratics establishes that a belt of country 2 to 10 km wide was covered by ice from the latitude of Bray to Coolgreany, then widening to about 16 km as far as Wexford town. In the Screen Hills, north of Wexford Harbour, retreat and decay of the Midlandian ice (from stage 3IS) has left a wonderfully fresh topography of morainic ridges, kettle holes, kames, and a great

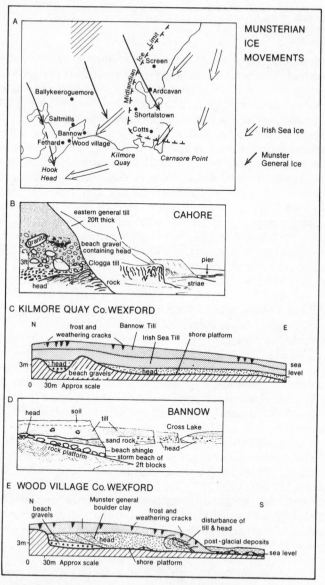

Fig. 6.7 Some aspects of the Quaternary geology of southeastern Co. Wexford, redrawn from diagrams by Mitchell and Synge.

mixture of limestone till and interbedded sands and gravels, much of it hardly weathered.

Southern Co. Wexford   In the extreme southeastern corner of Co. Wexford the investigations at Shortalstown have led to a re-assessment of the position of the outer limit of the Last (Midlandian) glaciation as already described (figs. 6.1 and 6.7). Colhoun and Mitchell (1971) have described the cryoturbation seen in the older (Munsterian) tills outside the Midlandian limit at several sites, including one a little north of Carnsore Point at Ballyhealy, also at Saltmills and Wood Village near Fethard (Mitchell 1962). The degree of cryoturbation observed in the older drifts, including deep cracking, erection of pebbles, and churning to a depth of 2 to 3 m, has never been observed in tills of Midlandian age. Mitchell (1971) has recorded numerous pingos in the older drift terrain, but only spasmodically within the Midlandian limit, and Forth Mountain with its tors and blockfields stood as a nunatak above the last glaciation ice. Cliff sections at a number of important coastal sites are shown in fig. 7.4 (p.188). Along the coast between Fethard and Carnsore Point drift sections show none of the fresh morphological features found on the Midlandian drifts, and the contrast with the Screen Hills topography is remarkable. Likewise many of these sections in the Munsterian drifts display evidence of deep disturbance by frost action, including pingos, and considerable weathering. It is particularly interesting that the soil survey of Co. Wexford (Gardiner and Ryan 1964) contributed considerably to our knowledge, by showing that the same soil associations occurred north and south of Wexford Harbour, within the Midlandian ice-limit, and that quite different soil associations occurred in the Munsterian drifts.

A basin in Midlandian glacial deposits at Coolteen contains late-glacial deposits (dated at 12 020 years B.P.), as do kettle holes at Curracloe (Mitchell 1951), and at Ballynaclash mud in a kettle hole gave a 14C date of 11 050 years B.P. (McAulay and Watts 1961). Similarly, the interglacial deposits at Shortalstown are sealed by late-glacial mud dated to 12 150 years B.P., and containing the remains of the Giant Irish Deer (*Cervus giganteus*), a common fossil animal of Irish late-glacial hollows (Mitchell and Parkes 1949).

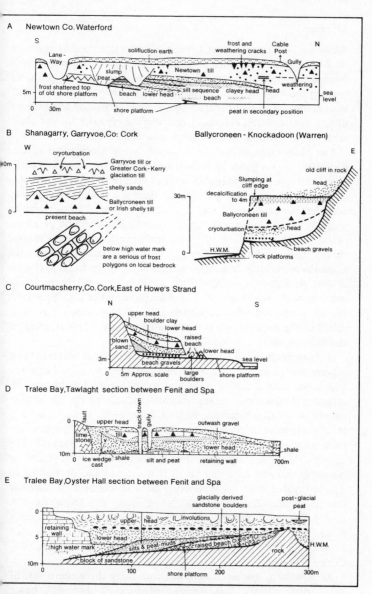

Fig. 6.8 Schematic cross-sections to illustrate some aspects of the Quaternary geology at selected sites in counties Waterford, Cork and Kerry, redrawn from diagrams by Farrington, Mitchell, Synge, and Watts.

Munster

Along the southeastern coast, between Kilmore Quay and Dargarvan, the Munsterian General ice pushed southwards and southeastwards from within Ireland, crossing the present coastline and probably meeting the Irish Sea ice. The latter ice had swept westwards after leaving the confines of the Irish Sea, moving across the southern tip of Co. Wexford and eventually reaching the type-site for the older Irish Sea shelly till at Ballycroneen, near Cork (see fig. 6.2, p.121).

The Munsterian inland ice deposited several distinct till facies, depending largely upon the underlying bedrock type, but all of them quite different from the Irish Sea till in texture and erratic content. At Newtown, on the Barrow estuary, there is a thick till containing considerable amounts of Old Red sandstone and Leinster granite and overlying a Gortian peat layer (fig. 6.8). Elsewhere till of inland origin contains mainly local rocks, some chert and some Carboniferous limestone boulders. This grey of buff-coloured Ballyvoyle till is usually completely decalcified and severely cryoturbated to depths of 3m. At Kilbeg, Co. Waterford, Gortian peat-mud is probably overlain by the Ballyvoyle till. In the northern part of Co. Waterford the till contains more Carboniferous limestone, and it has been called the Mothel till (Watts 1959), whereas Synge (1964) refers to the Bannow till in Co. Wexford.

It seems likely that the Ballyvoyle, Bannow, Mothel and Newtown tills are related members making up a formation of glacial deposits laid down during the advance and retreat of an ice-sheet which extended southwards from the Irish Midlands during the Munsterian glaciation. In Co. Wexford, between Kilmore Quay and Bannow, the superimposition of the till members indicates that the Bannow till is somewhat younger than the Ballycroneen (Irish Sea) till. The Bannow till probably evolved from a mixture of inland (Mothel or Newtown till) and Irish Sea till. It is weathered and decalcified to depths greater than 2.4 m, cryoturbated to about 2 m, and the erratics include some large boulders of Leinster granite (see fig. 6.7, p.137). The southward-moving inland ice may have been sufficiently powerful to hold the westward-moving Irish Sea ice off the coast between Kilmore Quay, Co. Wexford, and Ballycotton Bay, Co. Cork.

At Newtown, extensive sections occur along the western side of the Barrow estuary. The abrasion platform is seen with beach gravels cemented to it, with Gortian peat resting on the pebbles, and overlain by Newtown till. Considerable slumping obscures the stratigraphy, but at one place the rock-platform has been severely frost shattered, with considerable cryoturbation of the fragments, the whole overlain by till, which is also disturbed at the surface to a depth of 2 to 3 m (fig. 6.8).

The Greater Cork-Kerry glaciation The ice-sheet which extended from the Cork/Kerry mountains to cover eastern Co. Cork deposited extensive spreads of rubbly boulder clay, which have been heavily weathered and smoothed, and they now lack any fresh topographical forms. The drift-sheet is discontinuous because in many places it has been removed by later erosion. The ice passed off the south coast, extending as far east as the mouth of the River Blackwater at Youghal, while in the north it pushed towards Mallow and the head of Dingle Bay (see fig. 6.2, p.121; Farrington 1954). Both the Munster General ice from the Irish Midlands and the Irish Sea ice advanced into Co. Cork. In places the stratigraphical succession reveals the chronology of the various advances, notably the relationship between the Cork/Kerry ice-cap and the westward advancing Irish Sea ice, which is known to have reached Cork Harbour at Ballycroneen (table 6.3).

At Shanagarry-Garryvoe, on Ballycotton Bay, the Garryvoe till of western origin overlies shelly outwash-sand derived from the Lower Irish Sea or Ballycroneen till (Farrington 1966b). The latter is still remarkably consistent in composition, being composed of a stiff calcareous clay containing Carboniferous limestone and far-travelled erratics, these including Scottish rocks such as Ailsa Craig microgranite. Immediately underlying the Irish Sea till at Garryvoe is an area of frost-shattered limestone bedrock, with the angular debris arranged in polygons and stripes (Farrington 1966b) indicating the severity of the climate as the large ice-sheets were extending across the country. The Irish Sea till has an undulating, dissected surface on which rests up to 6 m of fine sands and silts containing shell fragments derived from the underlying till. These sands must represent outwash deposits laid down by the water responsible for the erosion of the surface of the underlying till. Subsequently, the eastward advance of the Cork-Kerry ice sealed the shelly sands below a variable thickness of rubbly (Garryvoe) till containing mostly local rock types (see fig. 6.8, p.139).

Table 6.3 Tentative chronology of events during the late-Pleistocene in Munster

Post-glacial	Littleton muds and peats, at Littleton, Co. Tipperary Low raised-beaches in Bantry Bay and the Shannon Estuary?		
Midlandian glacial period	*Co. Kerry* *Pollen Zone II* Finglas mud (base at 11950 years B.P.) Cirque glaciers *Lesser Dingle Ice Advances* marked by 'fresh' moraines at Anascaul Lough, Ballinloghig and Cruttia. Moraines in Behy and Cummeragh valleys.	*Co. Cork* Whiddy Island mud at −55m. Cirque glaciers *Lesser Cork-Kerry Ice Advances* Killumney and associated moraines	
Ipswichian interglacial	No type site is known in Munster		
Munsterian glacial period	*Co. Kerry* *Dingle Peninsula* Carrigaha tills Finglas gravel fan Feohanagh Upper till Feohanagh pollen sands Feohanagh Lower till Coose head – – – – – – – – – – Ballymore till	*Fenit and Spa* Upper head Tawlaght till and outwash gravels Lower head	*Co, Cork* *Greater Cork-Kerry Ice Advances* Garryvoe till – – – – – – – – – – Knockadoon Upper head Irish Sea Ice Advance Ballycroneen shelly till Ballycroneen Lower head
Gortian Interglacial	Dingle raised-beach?	Peats and silts Raised-beach	Raised-beach
	Co. Limerick: Baggotstown mud and Kildromin mud		
	At the coast the raised-beaches believed to be of Gortian age usually rest upon a well-developed abrasion platform, and the beach gravels frequently contain erratic pebbles.		

1 Events placed side-by-side need not have occurred at exactly the same time
2 A dashed line between adjacent deposits/events indicates that there is no stratigraphical relationship known, and different locations may be implied.

Source: Based upon the work of Lewis, Farrington, Mitchell and Synge.

In Co. Kerry, Lewis (1967, 1974) has identified moraines believed to indicate the outer limits of mountain glaciers during the Munsterian Greater Cork-Kerry glaciation in the Behy valley. These valley moraines are believed to be equivalent in time to the Tawlaght till, deposited by glaciers from the Kerry Mountains, which Mitchell (1970a) identified as overlying (together with Upper and Lower Head) the Ballymakegogue Gortian peats and silts in cliff-sections between Fenit and Spa, Co. Kerry (see fig. 6.9, p.139 and table 6.2). The Gortian deposits rest upon raised-beach gravels, and all overlie a wave-abraded rock-platform (Mitchell *et al.* 1973).

The Lesser Cork-Kerry glaciation   The limits of the Lesser Cork-Kerry ice-cap are indicated in figs. 6.1 and 6.2 (pp.117 and 121) where the ice is shown to cross the coast east of Mizen Head and Cape Clear. Its eastern limit in the Lee valley is marked by a huge gravel moraine at Killumney (Farrington 1959), although the limit has been extended farther to the east by Synge. The ice extended north to Killarney, encircling but not covering Macgillycuddy's Reeks, and a lobe may have just reached the head of Dingle Bay (Farrington 1954), although Lewis (1967) argues that ice was confined to the upper valleys at this stage, with fresh moraines of Coomasaharn in the Behy valley marking the outer limit. A moraine at Waterville, Co. Kerry, is believed to mark the western limit of ice which originated in the Kenmare valley and moved northwards through the Iveragh Mountains (Bryant 1968). Substantial corrie development has taken place in the Cork-Kerry Mountains, particularly fine examples occurring near Killarney on Mangerton Mountain, in Macgillycuddy's Reeks, and in the Caha Mountains of the Beara peninsula.

Co. Limerick and Co. Tipperary   The southern limit of the Midlandian ice is defined in many places at stage A by a clearly marked morainic belt (figs. 6.2, p.121 and 6.9). Its position in southeastern Co. Limerick is marked by the southern limit of fresh limestone till against the Old Red sandstone hills forming the foothills of the Galty Mountains. It is particularly well-marked as hummocky kame moraine in the Ballylanders Basin (Ballylanders Moraine), and in the Glen of Aherlow, where drainage channels and ice-marginal deltas testify to various stages of subsequent ice disso- lution (Synge 1970). Near the type-site for the terminal moraine of

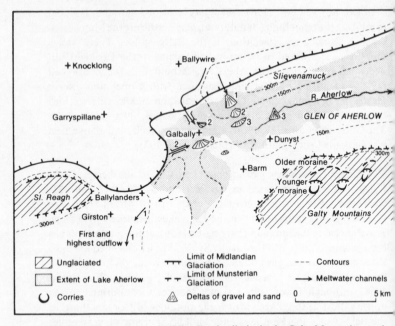

Fig. 6.9   Munsterian and Midlandian ice-limits in the Galty Mountains, and the maximum development of Glacial Lake Aherlow during the Midlandian, redrawn from a map by Synge. The numbered deltas (1, 2, 3) indicate successively lower levels of Lake Aherlow as ice retreat occurred.

the inland Irish ice at Ballylanders, Co. Limerick, the southern limit of fresh limestone till can be observed banked against the north-facing slope of the Old Red sandstone foothills of the Galty range. The ice exploited any low cols or gaps in these foothills, and succeeded both in penetrating the Ballylanders basin and in closing the eastern end of the Glen of Aherlow. Subglacial and ice-marginal drainage channels and deltas indicate that for a time a proglacial lake was impounded in the Glen of Aherlow, the various deltas (Corderry delta at 168 m, Galbally delta at 122 m, and the Lissvarrinane delta at 90 m) recording the successively falling levels of the lake as ice dissolution occurred (fig. 6.9). At Fedamore, Co. Limerick, further morainic ridges mark the outer limit of drumlins in this area. The drift limit reaches only about 230 m near Ballylanders, whereas on the western flank of the Galty Mountains glacial erratics associated with the 'older drift' reach 360 m. Synge has also demonstrated that independent corrie glaciations on the north-facing slopes of the Galty Mountains can be divided into a lower (335 m) 'older' series and a higher (492 m) 'newer' series at Lough Muskry. The 'older' series is marked by much degraded outermost moraines, while the 'newer' consists of two arcs of fresh block moraine; the latter can perhaps be correlated with the Athdown and late-glacial (Lough Nahanagan) stages of corrie glaciation in the Wicklow Mountains.

The Munsterian glacial period is considered to be represented by till resting on Gortian organic deposits at Baggotstown, and Kildromin, Co. Limerick (Watts 1964), and by till found at Ballyorgan and Athea, Co. Limerick. Both Baggotstown and Kildromin lie inside the terminal moraine (Ballylanders Moraine) which constitutes the southern limit of a substantial sheet of calcareous till, whereas the older tills have a considerable sandstone component. At Littleton, Co. Tipperary, Mitchell (1965) has established a chronology for the late- and postglacial vegetation hitory based upon a series of muds and peats recovered by boring (fig. 6.25, p.177).

Connacht

During the Munsterian Glacial period an ice-cap centred over the Connemara Mountains was of sufficient size to force ice southwards to the Mullaghareirk Mountains of Cork and Limerick; to Water-grasshill, Co. Cork; and to Thomastown, Co. Kilkenny, in all of which localities erratics of Galway granite are recorded (Charlesworth 1953). Limestone drift was also carried to 183 m on the northern side of the Dingle peninsula, and it is probable that contact was made between the ice moving southwards and glaciers moving northwards into Tralee Bay from cirques in the Dingle peninsula.

In counties Clare, Galway, Mayo and Sligo, but notably in northwestern Mayo, there are considerable areas of weathered and deeply cryoturbated till. These drifts do not form any fresh glacial landforms, such as eskers, kames or drumlins. Synge (1968, 1969) refers to several facies of 'older drift' (Erris, Killadoon and Belderg tills) in northwestern Connacht, although the Killadoon till is now regarded as of possible Midlandian age. In Erris the ice flowed generally northwestwards. In Murrisk ice streaming westwards from the centre of Ireland interrupted the radial outflow of mountain glaciers from an ice-axis extending from the Sheeffry Hills to the Maumturk Mountains across Killary Harbour. Ice probably failed to surmount the summits of Mweelrea (820 m) and Croagh Patrick (773 m) for an erratic limit is recorded at 580 m on the former peak. Synge also shows the ice to have had a gradient of about 7 m per km

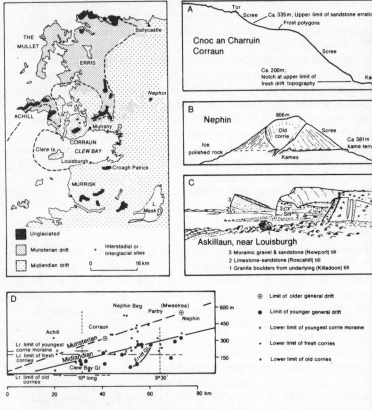

Fig. 6.10 Glacial drifts, and possible unglaciated areas, in northwestern Co. Mayo.

(A) Upper limit of sandstone erratics (= Munsterian?) and upper limit of fresh kames (= Midlandian and Drumlin Readvance limit?).

(B) Upper limit of fresh kames against Nephin, and an old corrie.

(C) Three till members exposed in cliff sections near Louisburgh.

(D) Height-range diagram of Munsterian and Midlandian glacial features between Nephin and Achill Island.

Figure redrawn from diagrams by Synge.

from Mweelrea towards Corraun and Achill Island (fig. 6.10). In Erris the ice flowed generally northwestwards but may have failed to cross all the high ground above 183 m along the coast between Belderg Harbour and Broad Haven. Between Belderg and Ballycastle, Co. Mayo, small pockets of a chocolate-coloured Belderg shelly till are recorded, and these may have come from the north across Donegal Bay. It is conceivable that it is the same till facies as the shelly till of Scottish origin found in Lough Foyle and at Malin Beg (see fig. 6.2, p.121). The tills derived from inland ice (Erris and Killadoon tills) are quite different, being largely composed of local gneiss and quartzites, or shale and sandstone.

Synge (1968) has recorded three sites in west Co. Mayo where organic material is seen between the 'older drift' and 'newer drift'. He refers to two of these deposits as the 'Louisburgh cool phase' (Old Head deposit) and the 'Corraun interglacial' (Cartron River deposit); the third deposit on the north coast of Co. Mayo (at Glendossera) may not be *in situ*. Pollen analysis is unable to confirm the interglacial status of the deposits without reservation, although such a status seems most likely in the case of the Cartron River deposits lying upon 'older' Erris till, and overlain by the Ballycastle-Mulrany Moraine (see table 6.4, p.149). The latter apparently represents the limit of the Midlandian glaciation (stage C, fig. 6.2) in western Ireland. The type-site for the Irish Gortian interglacial is at Boleyneendorrish, near Gort, Co. Galway, where two till members bury the organic mud, the stratigraphy being revealed in a stream section.

The western limit of the fresh drift topography in west Co. Mayo has been re-mapped by Synge (1963b, 1968), and he indicates a general ice-gradient for the Clew Bay glacier of about 5 m per km (fig. 6.10). The Ballycastle-Mulrany Moraine corresponds closely to the drumlin limit in Clew Bay; ice reached an upper limit of about 381 m around Nephin Beg Mountain, but only 198 m on the southern side of Corraun. Similarly, the ice-limit on the northern flank of Croagh Patrick lies between 275 and 305 m, forming a massive drift terrace separating quartzite scree above from fresh kames below. Two till facies (Newport sandstone and Roscahill limestone) were recognised by Synge (1968) in the Clew Bay drumlins, but neither exhibits any significant weathering features. At Askillaun, near Louisburgh, the presence of copious granite boulders on the foreshore suggests the Killadoon till is concealed by the later two facies of the Midlandian tills (see fig. 6.10, p.146).

In Murrisk, Killary Harbour is undoubtedly an over-deepened glacial trough, and fresh drift appears to be present over most of Murrisk and the area to the south and southwest of Killary Harbour. Especially noteworthy is an area of fine morainic topography to the southeast of Leenaun and Culliagh Beg in the valley of the Joyce's River.

The independent corrie glaciation of the western mountains was partly contemporaneous with the maximum extent of the Midlandian ice. Not all of the sixty or so corries in western Co. Mayo are of the same age, for some are apparently ancient features partially filled with scree (Synge 1968) as, for example, on the northern slopes of Nephin Beg, on Nephin, and on Croagh Patrick, and there is an absence of attached morainic forms. Some very old and much degraded moraines are found in areas peripheral to the Munsterian ice, for example, in the most westerly corries on Achill Island. One such old moraine is truncated by the sea below its corrie of Lough Nakeeroge East. Four phases of a younger (probably Midlandian) corrie glaciation have been recognised on Achill, notably in Acorrymore corrie (Farrington 1953b). In the Nephin Beg mountains small glaciers from a number of corries, such as Corryloughaphuill, Lough Anaffrin and Lough Nambrackkeagh, were synchronous with the Midlandian glaciation maximum (= Ballycastle-Mulrany Moraine and drumlin limit). The corrie moraines and the limiting moraine of the general ice coalesce as one unit of deposition, for example at Doo Lough below Ben Gorm, and at Lough Nambrackkeagh. The same kind of relationship was observed in the Partry Mountains, between Murrisk and Lough Mask (Synge 1968). Thus it seems that during the Midlandian glaciation in western Co. Mayo the highland areas acted chiefly as barriers to westward moving ice from the Central Lowland, and local glaciation was limited in most cases to small corrie glaciers. The evidence suggests that the chronology shown in table 6.4 may be correct.

Drumlins are particularly well displayed around the head of Clew Bay and were associated with the actively advancing Midlandian ice, but subsequently the 'retreat' stages were not marked by extensive morainic ridges or esker development, for these forms are largely absent.

In Co. Sligo there is less evidence available of the older drifts, but during the Midlandian glaciation the summits of the Ox Mountains escaped inundation, as did two small areas on Magho Mountain

Table 6.4

	General	Local Mountain
Midlandian glaciation	Ballycastle-Mulrany End-Moraine Newport till Roscahill till Killadoon till	Four stages of young corrie glacier development on Achill-Acorrymore moraines Anaffrin moraines in the Nephin Beg mountains
Last Interglacial	Cartron River deposits in Corraun?	
Munsterian glaciation	Belderg shelly till Erris till Gort Upper Solifluction gravel	Several stages of old corrie glacier development on Achill-Nakerroge moraines
Gortian Interglacial	Gort polleniferous mud	

(>300 m) and on Cuilcagh Mountain (>500 m), where no glacial striae were recorded, but where periglacial features abound (shattered rock, block screes, and pillar-shaped tors up to 14 m high). Small corrie glaciers were present in the uplands along the Co. Sligo-Co. Leitrim border, and landslips and extensive screes are present only above the prominent 'trim-line' of the general ice. The ice swept westwards to cross the present coastline, probably during the late or Drumlin Readvance stage of the Midlandian glaciation. Drumlins stream westwards and northwestwards towards Sligo Bay from the Midlands, and westwards from the Lough Erne lowland on the northern side of the Cuilcagh Plateau Country. This movement westwards in the Lough Erne area has been substantiated by studies of striae, stone orientations in till, erratic carriage, and carbonate content of till samples (fig. 6.12). Chapman (1970) established that there was no evidence of the carriage of Donegal schist erratics southeastwards to Lower Lough Erne from Co. Donegal. The Midlandian ice, which flowed generally westwards towards Donegal Bay, appears to have had an early source area in western Co. Tyrone, but a later source in an extensive ice accumulation which formed over the lowlands of eastern Co. Tyrone and the Lough Neagh basin.

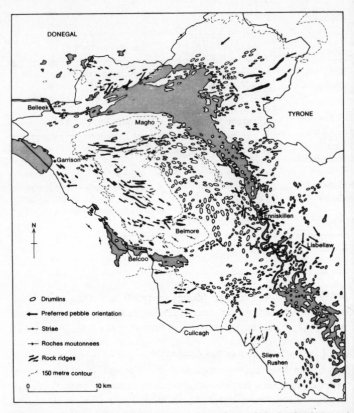

Fig. 6.11   Some aspects of the Quaternary geology of the lowland around Upper and Lower Lough Erne, and the adjacent upland areas in Co. Fermanagh. The preferred pebble orientation of stones in the drumlin-forming till indicates that north of Enniskillen there was a general westward movement of ice, while to the south of the town a soutward direction was important. Drumlin orientation and till fabrics were probably affected by the presence of high ground southwest of the lakes. The rock ridges are streamlined, ice-moulded ridges in bedrock. Figure redrawn from diagrams by Chapman.

Fig. 6.12   A tentative chronology of events during the Late-Pleistocene in northwestern and northeastern Ulster. The approximate extent of the various ice masses is shown schematically, as well as the timing of certain marine transgressions. This figure is based upon the work of Charlesworth, Colhoun, Creighton, Dwerryhouse, Prior, Stephens, and Synge.

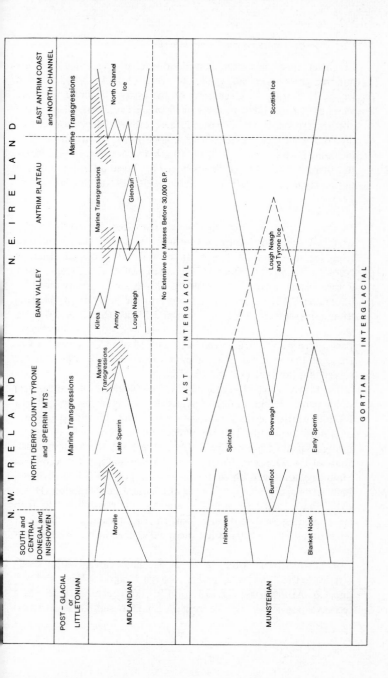

Ulster (with counties Louth and Meath)

In Ulster there has been considerable controversy for the last hundred years concerning the relative roles of the Irish and Scottish ice-sheets during the glaciations of the area. The rather extreme views of Kilroe (1888) in carrying Scottish ice across Ulster from northeast to southwest have been largely rejected. The importance of inland Irish ice-centres has been emphasised by various writers, including Dwerryhouse (1923) and Charlesworth (1924, 1939, 1963b). But there is no doubt that Scottish ice has played an important part during some glacial periods (Colhoun 1970, 1971a; Stephens *et al.* 1975).

Western Ulster   During the Munsterian Cold Period in north-western Ulster, northward-moving Irish ice crossed the north coast from ice-accumulation centres in central Co. Donegal (Blanket Nook glaciation of Colhoun 1970), and in central Co. Tyrone (Early Sperrin glaciation of Colhoun 1970). This expansion of ice may also have carried Tyrone igneous erratics into the heart of the Mourne Mountains, and to counties Louth and Meath (McCabe 1973), as well as to the Bann valley and the lower Roe valley at Bovevagh. Barnsmore granite erratics were also carried from southwestern Co. Donegal to the lower Roe valley because they are found in the succeeding Bovevagh till (fig. 6.12).

The next stage of glaciation involved an incursion of Scottish ice, which penetrated into Lough Foyle, extended as far as the northern foothills of the Sperrin Mountains, and accounted for the deposition of the Bovevagh till. This ice reached the southern end of Lough Swilly (Burnfoot till) and probably transgressed the western coast-lands of Co. Donegal, for the same type of shelly till is known at Malinbeg, and even on the northern coast of Co. Mayo between Ballycastle and Belderg, as well as at Glenulra, and Aghdoo (Hinch 1913; Synge 1968). There is of course also the possibility that Irish ice expanding westwards into Donegal Bay picked up submarine muds and shells and deposited shelly till at the various coastal sites mentioned.

In northeastern Ulster, ice from Scotland carried Ailsa Craig erratics to the Bann valley and the Lough Neagh lowland, and well into Co. Armagh (see fig. 6.1, p.117). Subsequently, there was a second expansion of Irish ice from the Lough Neagh lowland, across

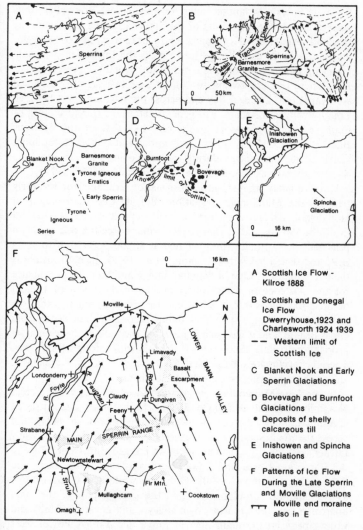

Fig. 6.13 Directions of ice movement in northwestern Ulster according to various authorities. Redrawn from maps by Colhoun.

the main Sperrin Mountain range, into the Upper Roe valley, thus accounting for the deposition of the Spincha till. At the same time there was expansion of ice from a centre, or centres, in Co. Donegal, and in the north this allowed ice to cross the Inishowen peninsula (carrying Barnesmore granite erratics) to beyond the present coast-line. There is every reason to believe that this ice also extended westwards into Donegal Bay and across the Atlantic coast between Lough Swilly and Ardara. The general pattern of this Inishowen or Spincha glaciation was probably very like that of the previous Blanket Nook or Early Sperrin glaciation, and similar to the more restricted ice movements during the succeeding Moville and Late Sperrin glaciations (fig. 6.12, fig. 6.13) (Colhoun 1970, 1971a).

Unfortunately, there is little information available for the early part of the Midlandian glaciation. But in Co. Fermanagh, at Hollybrook Bridge, a distinctive head deposit is overlain by till (McCabe 1969a), and in Derryvree townland (see fig. 6.2, p.121) a peat bed with a tundra flora is dated at 30 500 ± 1170/1030 years B.P. and sealed below till (Colhoun *et al.* 1972). In both instances the overlying till is Late Midlandian in age and there is no evidence for the existence of extensive ice-masses in the north of Ireland before 30 000 years B.P. of sufficient size to occupy much of the area to depths exceeding 300 m. Such dating accords with evidence from Great Britain, and it has been argued that the bulk of the ice-masses probably post-date 25 000 years B.P., following a long period when a periglacial climate and tundra landscape existed in Co. Fermanagh, and probably throughout much of the island.

In the Late Midlandian, an ice-sheet developed over the Lough Neagh lowland and central Co. Tyrone. From these main centres of accumulation ice moved northwards along the Bann valley and across the main Sperrin range, exceeding 700 m in elevation, while farther west ice from the Omagh basin merged with Donegal ice in the Barnesmore area and flowed northwards along the Derg and Finn valleys (Colhoun 1970). Only the western margins of Co. Tyrone and Co. Londonderry (i.e., the western part of the main Sperrin range, the Mourne-Foyle valleys, and an area of hills and valleys between the main Sperrin range and Lough Foyle) were influenced by ice from Donegal. The distribution of Barnesmore granite erratics (fig. 6.13) indicates that Donegal ice did not penetrate eastwards to the Lough Neagh lowland during the Late Midlandian as Dwerryhouse (1923) and Charlesworth (1939, 1963b) suggested. Furthermore, some ice from the Barnesmore area flowed southwestwards to Donegal Bay, completing the erosion of the deep U-shaped Barnesmore Gap, and joining with other ice tongues to form a wide lobe in Donegal Bay. The western limit is indicated by moraines at Shalwy and Muckros Head, and by the outer limit of drumlins extending from the head of Donegal Bay to beyond Killybegs, and from Ardara to Maghera and across the Loughros peninsula to the north-facing sea cliffs of the Slievetooey Mountains. The same limit occurs as a well-marked morainic line at Ballycastle, Co. Mayo, where it appears to mark the western extremity of fresh morphological drift features and associated till and fluvio-glacial deposits. Outside this line deeply weathered and soliflucted till is found in conjunction with extensive head deposits to form a subdued landscape lacking in kames and steep ice-contact slopes.

The Glengesh plateau in southwestern Co. Donegal lay beyond the limit of the ice carrying the Barnesmore granite erratics westwards, but it is considered to have supported a local ice-cap from which glaciers moved outwards in a radial pattern (Colhoun 1973). The ice-surface sloped steeply from above 600 m in the east, where the ice joined with the general ice along an arc between Ardara and Shalwy, to about 300 m, on Slieve League, 320 to 360 m on Slievetooey, but to below 200 m where the west-flowing outlet glaciers of Glenlough, Port and Glencolumbkille scoured and deepened their valleys and crossed the present coast. Several large nunataks remained, including Slievetooey, Slieve League, and an area between Leahan and Malin More. In all these extra-glacial areas extensive blockfields, screes, buttresses, and tors indicate the severity of the periglacial processes.

In Inishowen the Moville and Late-Sperrin glaciations carried ice to a limiting moraine (figs. 6.13 and 6.14), which has also been indicated as the outer limit of drumlins (Stephens and Synge 1965; Colhoun 1970). This Drumlin Readvance represents the outer limit of fresh morainic hummocks near Moville on Lough Foyle and in Lough Swilly at Dunaff, and it is believed to link to a similar moraine near Maghera, which forms the limit of drumlins near Ardara and on the Loughros peninsula. Severe periglacial processes have produced deeply disturbed till outside this limit in Inishowen and Fanad, where thick head deposits and extensive screes occur. There is no definite proof that Midlandian ice covered Inishowen and Fanad beyond the limits of the Drumlin Readvance, although this possibility requires further investigation.

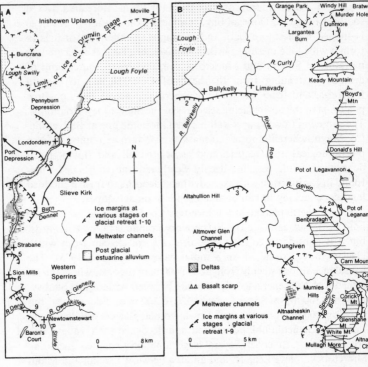

Fig. 6.14 Stages of retreat of the Midlandian ice in the Foyle, Derg and Roe valleys. Recession took place from north to south, and across the Sperrin ridges and valleys to the Omagh basin and Lough Neagh lowland. This figure redrawn from maps by Colhoun.

The Drumlin Readvance moraine is also partially bevelled in places by marine benches, and red marine clay can be seen to interdigitate with the morainic deposits, indicating the contemporaneous nature of the maximum ice advance and a marine transgression (Stephens and Synge 1965). Outside the morainic limit, which is not over-stepped by any later ice movement, magnificent storm shingle beach-bars extend up to 23 m at Malin Head, Fanad Head, and at Ballyhillin, and marine washing limits can be seen at many sites in Inishowen and Fanad (see fig. 7.9, p.202 and fig. 6.12).

The manner of the deglaciation in parts of northwestern Ulster has been worked out by Colhoun (1972), whose findings are at variance with those of Charlesworth (1924, 1963b). Colhoun considers that there was a southward withdrawal of the ice lobes in Lough Foyle and Lough Swilly from the Drumlin Readvance end moraine, and probably a contemporaneous withdrawal of Donegal ice from the western coastlands and Donegal Bay (fig. 6.14). Within Lough Foyle well-developed notches in drift at about 15 m occur at Drumskellan, and a magnificent 15-m marine terrace, pitted with kettle holes, is seen between Londonderry and Ballykelly, where the sediments have been disturbed subsequently by freeze-thaw processes to depths of 2 m. Subsequently, the ice-sheet margin retreated southwards, firstly on a continuous front in the Faughan and Roe valleys, where the ice became separated from ice in the Lower Bann valley by the plateau basalts forming the north Derry hills. Ice in Lough Swilly melted faster than the Foyle glacier, and meltwaters from the latter passed westwards through the large Port Lough depression. The later southward retreat of ice in all the northward sloping valleys, as well as in the Lower Bann valley, and the continuance of this withdrawal southwards beyond the watershed of the Sperrin Mountains, indicates conclusively that the main source areas for the northwestern ice-sheets lay to the south of the mountains. The main stages of retreat are indicated in fig. 6.14, and Colhoun (1972) has demonstrated convincingly that the northward direction of glacial meltwaters responsible for outwash sands and gravels, and the pattern of rock-cut meltwater channels, indicates free outlet of glacial drainage along the Foyle, Faughan and Roe valleys to Lough Foyle. No evidence could be found to support the existence during deglaciation of extensive and deep ice-dammed glacial lakes interconnected by meltwater channels, either to the north or to the south of the main Sperrin watershed. The main ice-sheet retreated southwestwards and westwards, exposing ridges and valleys (Glenelly, Glenlark and Coneglen), towards the centre of ice accumulation in the Omagh basin and central Co. Tyrone. Meltwater channels, moraines, eskers, and ice-contact deltaic deposits on the valley margins all fit to a pattern indicating synchronous deglaciation of this extensive system of valleys, tributary to the Owenkillew valley, and thence to the Mourne-Foyle system. In each area thick ice occupied all the valley floors until deglaciation was complete and there is no evidence of former glacial lake shorelines and the thick laminated lake deposits which would indicate the existence of either separate or interconnected large lake systems as suggested by Charlesworth (1924).

Similarly, to the west of the Mourne-Foyle valley, the ice-sheet contracted southwestwards along the Finn, Derg and Baron's Court valleys into Co. Donegal. There must have been more or less synchronous decay and easterly withdrawal of Donegal ice from the western coastlands into the mountainous heart of Co. Donegal.

Eastern Ulster   In northeastern Ireland the homogeneous nature of the glacial tills developed on the extensive Antrim basalts militates against sub-division in the absence of *in situ* interglacial deposits. Consequently it is difficult to make close correlation with events in northwestern Ulster, especially during the Munsterian Cold Period. However, there seems to be general agreement that Scottish ice thrust southwestwards across the area (figs 6.1, 6.12 and 6.13, pp.117, 151 and 153), to a limit traced from just west of Lough Neagh to the Mourne-Carlingford massifs and the coastal plain of counties of Louth and Meath (McCabe 1973; Stephens *et al.* 1975).

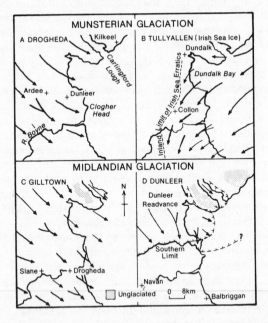

Fig. 6.15 Ice-flow directions during the Munsterian and Midlandian glacial periods in counties Louth and Meath. Redrawn from maps by McCabe.

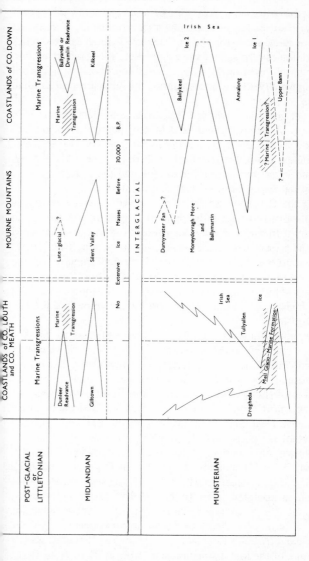

Fig. 6.16 A tentative chronology of events during the Late-Pleistocene in southeastern Ulster and adjacent areas of counties Louth and Meath. The approximate extent of the various ice masses is shown schematically, as well as the timing of certain marine transgressions. This figure is based upon the work of Charlesworth, Dwerryhouse, Hannon, Hill, McCabe, Stephens, and Synge.

This movement of Scottish ice was probably responsible for the widespread distribution of Ailsa Craig granite erratics, which cannot be explained adequately by known later ice movements. At the same time there is evidence for the carriage of Tyrone igneous series erratics to the eastern side of the Lough Neagh lowland, deep into Co. Down, and even on to parts of the Mourne and Carlingford mountains which were unglaciated during the subsequent Midlandian glaciation.

Although not strictly within Ulster, it is convenient to deal with early events in counties Louth and Meath, between Dundalk and Balbriggan, at this point. According to McCabe (1973) the greater part of these counties was affected by at least three general ice-sheets during the latter part of the Pleistocene and there were no local centres of ice dispersion (figs. 6.15 and 6.16).

During the Drogheda glaciation ice moved generally eastwards and southeastwards across the area from centres of dispersal in north-central Ireland and the Lough Neagh lowland, and extended beyond the present Irish Sea coastline. Striae trending from north-west to southeast have been recorded on the summits of the Carlingford Mountains, sometimes associated with weathered drift, and always above the uppermost limit of the Midlandian glaciation ice, which encircled both the Carlingford and Mourne massifs (see fig. 6.20, p.166).

The Drogheda till lacks Ailsa Craig microgranite and marine shell fragments, but contains calcareous pebbles and flints, which cannot have been derived from the overlying calcareous Irish Sea till (of Scottish origin). These pebbles were probably derived from chalk outcrops in the Lough Neagh lowland, and were carried southeast-wards by an early Munsterian ice-sheet. Such an expansion of Irish ice may have been contemporaneous with the Early Sperrin and Blanket Nook glaciations in northwestern Ulster (fig. 6.12 and 6.13).

In contrast, the overlying Irish Sea till (McCabe's Tullyallen glaciation, fig. 6.16) contains abundant shell fragments and Ailsa Craig microgranite. Separating the two tills is the Mell glacio-marine formation, a complex series of clays, silts, sands and gravels, recorded at Tullyallen quarry near Drogheda. These sediments appear to have been deposited penecontemporaneously as the Drogheda and Irish Sea ice-masses advanced. The contained marine arctic fauna of the Mell formation silts, lying at 29 to 40 m O.D., suggests accumulation in water depths of 90 to 110 m in a proglacial environment, with icebergs dropping erratics into the marine silts. The Drogheda ice must have acted as an important constraint to any extensive inland movement of the Irish Sea ice during the maximum of the Tullyallen glaciation. If the tripartite series of deposits at Tullyallen is all *in situ*, as we have every reason to believe, then isostatic depression must have occurred to permit seawater access to

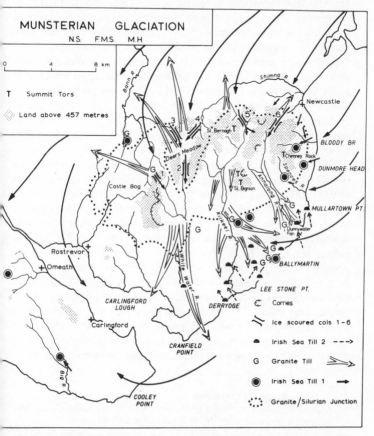

Fig. 6.17 Directions of ice movement in southeastern Co. Down, and a part of Co. Louth during the Munsterian glacial period. The six ice-scoured cols are: (1) Windy Gap, (2) Slieve Muck-Pigeon Rock col, (3) Spelga, (4) Ott Mountain col, (5) Hares Gap, (6) Glen River col. This map is based upon the work of Hannon, Stephens and Synge.

this site (Colhoun and McCabe 1973; Stephens *et al* 1975). Unfortunately, it is not yet possible to determine the position of world sea-level at this time, but if it had been at approximately present day level, then isostatic depression of the order of at least 150 m would be indicated (Synge 1977).

There was subsequently a major expansion of local ice-caps in the Mourne Mountains (figs 6.16 and 6.17). These accounted for the widespread distribution of huge quantities of granite boulders to situations far beyond the confines of the mountain valleys, and even beyond the present coastline to the southeast of Ballymartin and Kilkeel. Some large corries were initiated during this Moneydorragh or Ballymartin glaciation (e.g., on the western side of the Whitewater valley from Attical). The dispersal of ice from centres of accumulation over the upper Kilkeel valley (or Dorn area), and over the Spelga or Deer's Meadow area, was on such a scale as to imply that considerable withdrawal had taken place of both Scottish (Irish Sea) ice and Irish ice from the Lough Neagh area.

There was a minor readvance of the Scottish or Irish Sea ice (Ballykeel Readvance) across the line of the present coast to altitudes of 93 to 137 m O.D., where a number of broad, rounded ridges of drift mark the morainic limit a few kilometres inland of Ballymartin. In the Annalong valley the Dunnywater fan lies where the river discharges from its confined mountain course onto the Mourne coastal plain (fig. 6.17). The boulders and coarse, poorly sorted gravels probably represent a distinct phase of alluviation, and perhaps even a slight readvance of part of the mountain (Moneydorragh) ice-cap, discharging through the Annalong valley.

The three older till members and associated glacio-fluvial sediments can be seen in the cliff-sections at Ballymartin Quay, the type-site, although the Annalong till (Irish Sea or Scottish ice) is best displayed in Brackenagh townland, near the entrance to the Silent

---

Fig. 6.18 Ballymartin, Co. Down.

(A) Schematic section to show the stratigraphical arrangement of tills and certain ice-limits between the mountain edge and the coast.

(B) Simplified coast section north and south of the lane and slipway access to the beach.

(C) and (D) Valley cross-sections in the Silent and Annalong valleys showing the buried rock floors below glacial and fluvio-glacial sediments.

Figures (A) and (B) are redrawn from diagrams by Hannon and Stephens.

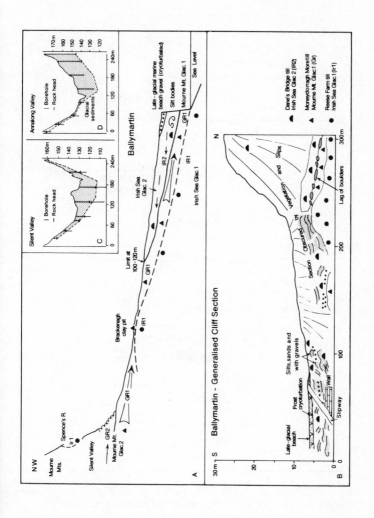

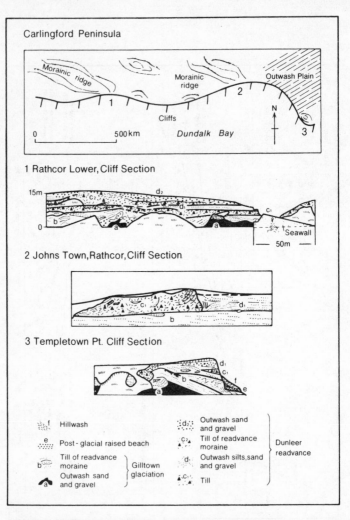

Fig. 6.19 Sketch map to show the location of the cliff sections between Rathcor and Templetown Point, Carlingford Peninsula, Co. Louth.

(1), (2) and (3): Cliff sections at Rathcor, Johns Town and Templetown Point. A series of small push-moraines (1st and 2nd Readvances) are exposed in the cliff sections, thus giving rise to a series of multiple till and fluvio-glacial sand and gravel members, each of which can be related to a particular morainic ridge. An outwash plain leading to the northeast is valley or upper Kilkeel River valley. This till, with a high silt-clay content, was quarried for 'puddle clay' during the building of the Silent valley dam (fig. 6.18).

The main advance of the succeeding Midlandian ice-sheets in counties Louth and Meath has been termed the Gilltown glaciation by McCabe (1973, and fig. 6.15C). This glacial advance was probably equivalent to the Kilkeel glaciation in southern Co. Down and the main expansion of the Lough Neagh and North Channel ice-sheet in northern Co. Down and in Co. Antrim. There are four till members making up the Gilltown till formation, which is considered to be stratigraphically equivalent to the uppermost till sheet in the Dublin area, and to the till sheets making up the extensive till plains in Central Ireland as far as the Ballylanders or Tipperary End Moraine (figs. 6.1 and 6.2, pp. 117 and 121). During the northward retreat of the Midlandian ice from the Dublin area, numerous minor pauses and slight readvances of the ice-front occurred; these are marked by numerous distinct morainic ridges between Dublin and Dundalk (McCabe 1972; Synge 1977); (fig. 6.19).

During the Kilkeel glaciation ice from the Lough Neagh lowland, together with some ice from southern Co. Antrim and the North Channel, moved to encircle the Mourne and Carlingford mountains reaching a maximum height of 427 m (Stephens *et al.* 1975). The limiting moraines at Kilkeel and at Dunmore Head suggest that a considerable area of the Mourne Plain remained unglaciated. There was no massive invasion of these mountains, above about 400 m, by extraneous ice, but within the Mournes three small separate valley glaciers formed in the Whitewater valley, Silent valley, and in the Annalong valley. None of these local mountain glaciers was able to make contact with the encircling general ice at its maximum extent (fig. 6.20).

The strong southerly flow of ice across Co. Down was matched by a general northward flow from the Lough Neagh lowland and the Belfast Hills, along the Bann and Main valleys, and along the east coast of Co. Antrim, where North Channel ice was probably also involved (figs. 6.21 and 6.22). The ice extended beyond the present northern coastline and in the west was probably co-existent with the

truncated by late-glacial shingle-bars near White's Town. These diagrams are based upon investigations by Synge, Stephens, and McCabe, and redrawn from a diagram by Synge.

ice which thrust northwestwards across the Sperrin Mountains. In the northeastern corner of Co. Antrim, in the Carey valley (fig. 6.21), a complex suite of deposits accumulated as ice-pressure waxed and waned from several directions. The stratigraphical arrangement of the tills and glacio-fluvial sands and gravels, often displaying distinct terraced forms, demonstrates ice movement from the south,

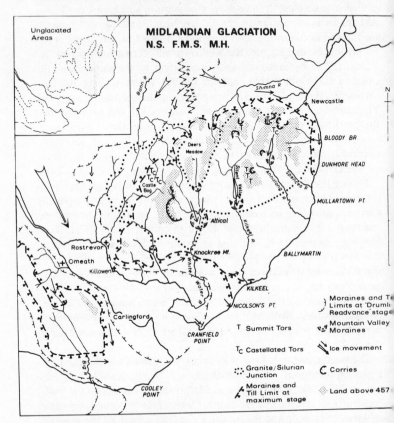

Fig. 6.20 Directions of ice movement and some ice-limits in southeastern Co. Down, and a part of Co. Louth, during the Midlandian glacial period. Some of the corries shown were initiated during the Munsterian (e.g. the large corrie lying to the west of the upper White Water River) and re-occupied by ice at this stage. This map is based upon the work of Hannon, McCabe, Stephens, and Synge.

east and northeast, all during the Late Midlandian glaciation. The last thrust of Scottish ice into the area from the northeast brought about the deposition of the Carboniferous-rich sediments in the Carey valley, and blocked this and other north-trending valleys by a massive morainic complex with multiple ridges which forms part of the Ballycastle-Armoy-Ballymoney Moraine. Extensive peat bogs occupy much of the low ground between the morainic ridges as, for example, at Gary Bog.

There is no evidence to support the views of Dwerryhouse (1923) and Charlesworth (1939, 1963b, 1973) that during the Midlandian Cold Period Scottish ice once again crossed the Antrim Plateau to join with Irish ice in the Lower Bann valley and the Lough Neagh basin. There is no trace of the carriage of Triassic, Keuper, Liassic or Chalk erratics from the coastal outcrops westwards across the Antrim Plateau, nor is there a recognisable Scottish till with far-travelled erratics.

The most widespread glacial deposit is a compact grey-brown basaltic till containing only local material. The upper limit of the general ice was at 260 to 280 m, and Prior (1968) has convincingly demonstrated a northward carriage of various erratics, including Old Red sandstone, Cushendall porphyry and Cushendun red quartz-porphyry (fig. 6.21). The till is always thickest on south-facing slopes, where it may rise to 150 m above that achieved on north-facing slopes in the Antrim glens. There is no evidence in the Antrim glens of the former existence of extensive lake systems linked by sequences of overflow channels such as were postulated by Charlesworth (1939, 1963b). On the contrary, the absence of lake shorelines and lake deposits, and the probability that many of the meltwater channels were marginal or sub-marginal in character, suggests downwasting of stagnant ice. Charlesworth envisaged the 'Antrim Coastal Readvance' extending from the Armoy-Ballycastle Moraine, along the coast to Belfast Lough. But the absence of a separate till member, or members, associated with such a readvance, of Scottish erratics on the coastal slope, and of clearly connected lake systems in the Antrim glens and in the Lagan valley, seems to largely invalidate his arguments. Northward-moving, and later retreating Irish and North Channel ice, probably accounted for most of the features on the coastal slope, as outlined above, and Hill (1968) has effectively demonstrated that a readvance into Belfast Lough is unnecessary to account for the sequence of glacial deposits recorded there. In a limited area above Glendun and Glenaam there was a small plateau ice-cap, although this may have been much larger (figs. 6.12 and 6.21). Unfortunately, the morainic and out-wash deposits of the plateau-top ice nowhere interdigitate with the deposits of the general coastal ice, and consequently the precise timing of the various advances in relation to one another is not known. A fine kame moraine, now overlain by blanket peat, at Orra Lodge in upper Glendun marks the maximum extent of the plateau ice in that valley (Prior 1970).

In southeastern Ulster, and in adjacent areas of counties Louth and Meath, it is abundantly clear that towards the end of the Midlandian Cold Period there was an important readvance of the ice-sheet from the Lough Neagh lowland (fig. 6.16). This Drumlin Readvance carried ice into Co. Louth and Dundalk Bay (the Dunleer Readvance of McCabe 1972; 1973; fig. 6.15D), to the north of Carlingford Lough, and to Killough and Ardglass in eastern Co. Down. In coastal areas the limiting end-moraine complex of the readvance is contemporaneous with a late-glacial marine transgression, the shorelines of which are terminated by the moraines (see fig. 7.18, p.218), while the marine sediments interdigitate with the associated tills and fluvioglacial deposits in Inishowen and in counties Antrim, Down and Louth. The Drumlin Readvance Moraines appear to have excluded the maximum of the marine transgression from much of Sligo Bay, Donegal Bay, Lough Swilly, Lough Foyle, Carlingford Lough, and much of Dundalk Bay, probably at the same time, 18 000 to 17 000 years ago (Stephens and Synge 1966a, 1966b; Synge 1977; Stephens and McCabe 1977).

The Drumlins The limiting moraines of the Drumlin Readvance in Ulster are shown in figs. 6.13 (p.153); 6.15 (p.158) and 6.20 (p.166), and the distribution of the main areas of drumlins in Ireland is shown in fig. 6.2 (p.121). By far the greatest concentration of drumlins lies within an area extending east to west from Co. Down to Co. Fermanagh, Co. Sligo and southwestern Co. Donegal, while a considerable number are found in the Lower Bann valley, north of Lough Neagh. The drumlin swarms contain individual hills of many

Fig. 6.21 Directions of ice movement and some ice-limits in the north of Ireland during the Midlandian glacial period. The extent of the independent ice masses forming on the Antrim Plateau is still conjectural. This map is based upon the work of Creighton, Prior, Stephens, and Synge.

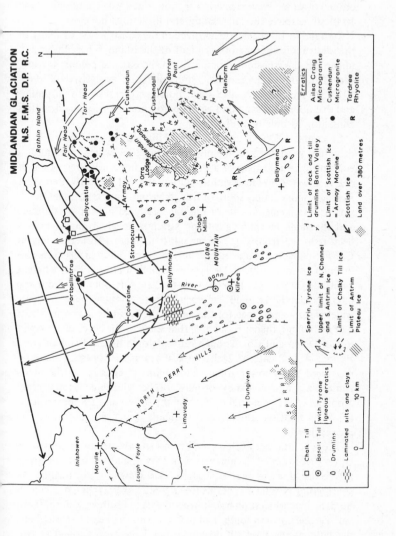

MIDLANDIAN GLACIATION
N.S. F.M.S. D.P. R.C.

Rathin Island

Fair Head
Torr Head
Cushendun
Cushendall
Garron Point
Glenarm

Ballycastle
Armoy
Orra Lodge
Glendun
Glenann

Ballymena

Portballintrae
Stranocum
Coleraine
Ballymoney
Clogh Mills
LONG MOUNTAIN
River Bann
Kilrea

NORTH DERRY HILLS
Dungiven
Limavady
SPERRINS

Inishowen
Moville
Lough Foyle

Erratics

▲ Ailsa Craig Microgranite
● Cushendun Microgranite
R Tardree Rhyolite

☐ Chalk Till
⊙ Basalt Till [with Tyrone igneous erratics]
0 Drumlins
〰 Laminated silts and clays

↗ Sperrin, Tyrone ice
⇓ Upper limit of N. Channel and S. Antrim ice
↓ Limit of Chalky Till ice
▨ Limit of Antrim Plateau ice

Limit of rock and till drumlins Bann Valley
Limit of Scottish ice 'Armoy Moraine'
← Scottish ice
▨ Land over 380 metres

0   10 km

N

shapes and sizes — they are rounded or elliptical in plan, sharp-crested or whale-backed in cross-section. The crests commonly stand 10 to 30 m above the intervening and ill-drained hollows.

Two major till sheets are involved in the construction of many of the till drumlins in the northern half of the island, where lower and upper tills have been differentiated on the basis of colour, texture, carbonate and erratic content, and till fabric analysis (Hill 1968, 1970, 1971). There are also small numbers of rock-cored drumlins and drumlins with sand and gravel cores, and occasionally drumlins consisting of only one till-type.

In counties Down and Antrim, Hill has shown that the till drumlins were formed mainly by deposition which produced a concentric arrangement of a thick layer of upper till over a core of older, lower till. Observations along the M1 Motorway in Co. Armagh, immediately south of Lough Neagh, have confirmed that a similar arrangement exists of one till 'plastered' on top of another, and the same morphological and stratigraphical arrangement has been reported from Co. Fermanagh (Chapman 1970). But only in Co. Fermanagh has an organic horizon been detected between two drumlin-forming tills, at Derryvree (Colhoun *et al.* 1972), where a 14C date 30 500 ± 1170/1030 years B.P. was obtained from the inter-bedded peaty mud. If such a date should prove to have widespread significance for the age of the two tills in the northern half of Ireland, then it is possible that some of the cores of lower till in Co. Down and Co. Armagh may pre-date the Midlandian Cold Period. A 14C date of 24 050 ± 650 years B.P. on shell material recovered from lower till in a drumlin at Glastry in the Ards peninsula, Co. Down, may, or may not, be valid evidence indicating a possible Late Midlandian age for the lower till. The proposition has been discussed already, but available evidence is unfortunately inconclusive.

There is little doubt that, as Hill has shown, drumlins consisting entirely of lower till in northern Co. Down, and the cores of lower till recorded elsewhere in eastern Co. Down, were formed by south-going North Channel ice (see fig. 6.22, p. 172). In Co. Armagh the principal orinetation becomes northeast to southwest, in southern Co. Tyrone north to south, and in Co. Fermanagh east to west. Thus the bulk of the Irish till drumlins may have been formed at two different periods, although it is evident that many more deep sections should be examined in detail to test the above hypothesis.

However, the internal composition of many of the drumlins, where there is a concentric arrangement of a thick layer of upper till over a core of lower till, confirms that they were formed largely by deposition, and not by erosion of an older drift by a later ice advance. There is also a close correspondence between the mean orientation of the till fabrics in the upper till and the orientation of the drumlin long axes, regardless of whether the drumlins have a core of lower till or not. In contrast, in the extreme northeastern corner of Co. Down, between Bangor and Newtownards, drumlins of lower till have an orientation corresponding approximately to the orientation of the lower till fabrics, which are unrelated to the direction of movement (north of west to south of east) of the later Irish ice to cross the area. Consequently, for this part of Co. Down, at least, the drumlins cannot have been formed by significant erosion and re-orientation of fabrics in either pre-existing drumlins or an older till plain (Hill 1971). Hill's investigations indicated that there was a tendency for changes in fabric orientation to occur in a consistent manner in a vertical profile taken through drumlins where both tills were recorded. The fabric orientations changed in such a way as to suggest regional shifts in ice flow direction, and that the drumlins were constructed by layer upon layer of till accretion through time. The till fabrics did not support a hypothesis that the drumlins might have been formed by the squeezing of plastic till into cavities in the base of the ice-sheet. But a certain degree of abrasion may take place as a till drumlin is formed, and there may have been some erosional moulding of the till as the over-riding ice fashioned the final streamlined shape. Where substantial rock cores occur in drumlins, examination suggests that in Co. Down and Co. Fermanagh substantial shattering of the rock by freeze-thaw processes prepared the rock surface before the moving ice completed the moulding of the rock surface by substantial erosion; 'tails' of shattered rock have been seen in till leading away in the direction of ice movement from the rock core.

In the areas of Co. Down and Co. Antrim examined by Hill (1968) four types of drumlin were identified according to their axial ratios (fig. 6.22; 1. 1.5:1; 2. 1.5:1 to 2:1; 3. 2:1 to 4:1; 4. 4:1). In general, the drumlin types dominate a series of parallel zones, which are elongated northwest to southeast, in the direction of movement of the Irish ice responsible for the deposition of the upper till. Consideration of drumlin density indicated that there was a tendency

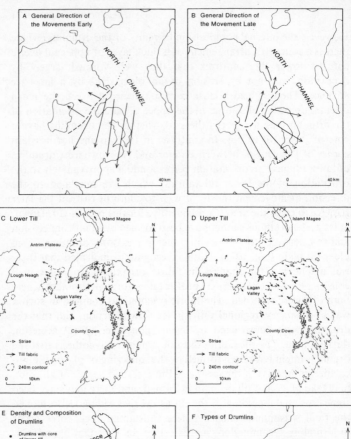

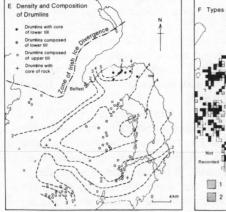

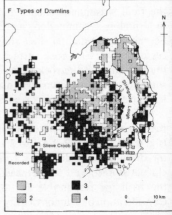

for high and low density zones to develop both parallel to and transverse to the direction of flow of the last Irish ice, a point confirmed by Vernon's (1966) work, especially in the southern part of the Ards peninsula and Strangford Lough where many islands consist either of single drumlins or groups of drumlins. There seems to be no relationship between these patterns and the topography and geological structure of the area, and explanation is perhaps to be sought in factors such as variation in the concentration of debris and variation in the rate of ice-flow within the ice-sheet. Occasionally the presence of obstacles, such as projecting knobs of bedrock or the concentrations of large boulders observed in various parts of eastern Co. Down, may have provided a node about which a drumlin has formed. But apparently non-random distribution of drumlins, and their concentration into zones parallel and transverse to the most recent ice movement, suggests that obstacles may not be an important factor.

Ice withdrawal and the Late-Glacial period in eastern Ulster In northeastern Co. Antrim Scottish ice advanced to the Ballycastle-Armoy-Ballymoney moraine (figs. 6.1, 6.2, and 6.21) following a limited withdrawal of ice occupying much of the Bann valley and North Channel. However, it now seems likely that Scottish and Bann valley ice were co-existent between Ballymoney and Armoy for a time, and consequently the complicated stratigraphical succession of deposits in this area resulted from the establishment of an extensive

---

Fig. 6.22 (A) and (B) General directions of ice movement during the Midlandian glacial period, redrawn from maps by Hill and Prior.

(C) Striae and till fabric measurements for the Lower Till involved in the construction of the drumlins.

(D) Striae and till fabric measurements for the Upper Till involved in the construction of the drumlins.

(E) Density and composition of the drumlins.

(F) A generalized map of types of drumlins classified according to their axial ratios. The axial ratio of all drumlins within each square of kilometre grid was examined. A drumlin 'type' was assigned to each square which contained 50% or more of drumlins of that particular type.

  (1) Axial ratio 1.5:1 ('circular').
  (2) Axial ratio 1.5:1 to 2:1 ('intermediate').
  (3) Axial ratio 2:1 to 4:1 ('lenticular').
  (4) Axial ratio 4:1 ('elongated ridges').

(C), (D), (E) and (F) are redrawn from maps by Hill.

medial morainic system which extended to the mouth of Lough Foyle (Creighton 1974).

Further withdrawal southwards of Bann valley ice exposed sites such as Canon's Lough, near Kilrea (Smith 1961), where late- and postglacial sediments are preserved in hollows in a kame-kettle moraine (figs. 6.22 and 6.23). Similarly, ice withdrawal from the Drumlin Readvance Moraine at Killough, Co. Down, permitted deposition of a sedimentary series in a partially enclosed depression (Stelfox *et al.* 1972). Thus the Lower Bann valley north of Kilrea, the Lagan valley at Belfast, and much of Co. Down were completely free of ice by the beginning of Pollen Zone II, about 12 000 years B.P., and the coastlands considerably earlier, possibly from about 18 000 years B.P.

Even if the Drumlin Readvances and attendant marine transgressions of the isostatically depressed areas in the north of Ireland were not exactly synchronous in time, it has proved possible to make some tentative assessment of the rates of ice withdrawal and marine activity (Mitchell 1972; Stephens *et al.* 1975). A summary is provided in chapter 7 where various positions of retreating ice-fronts are shown, together with tentative dating. The implication is that the sea flooded the isostatically depressed area to the limit provided by the various Irish and Scottish ice-sheets, and where in contact with land areas a series of shorelines, washing limits, and beach-bars was created. Following upon these late-glacial transgressions, isostatic recovery brought about an important withdrawal of the sea from existing coasts. Various coastal sites, such as Roddans Port, Co. Down (Morrison and Stephens 1965), the Glasgow area, Firth of Clyde, and southwestern Scotland (Gemmell 1973; Jardine 1964; 1971; Bishop and Dickson 1970; Peacock 1971) have yielded evidence for these late-glacial shorelines, and subsequent withdrawal of the sea. Recent studies of the postglacial elevated shorelines, and

---

Fig. 6.23 A schematic diagram to show the typical arrangement of sediments of Late- and Post-Glacial age found in numerous hollows (e.g. kettle holes and inter-drumlin hollows) throughout Ireland. Redrawn from a diagram by Smith.

Fig. 6.24 A pollen diagram from Long Range, Co. Kerry, showing a Late-Glacial pollen record. The percentages are based on the total pollen of woody plants and wind-pollinated herbs. The diagram includes only those pollen types that occur in substantial quantity. Redrawn from a diagram by Watts.

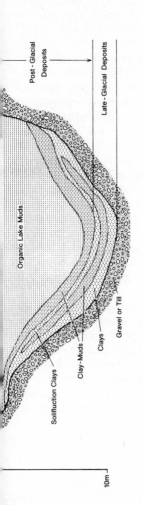

Fig. 6.23

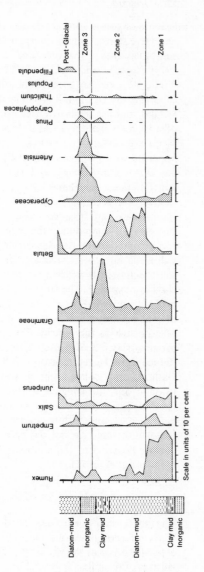

Fig. 6.24

beaches or coasts adjacent to the North Channel and northern Irish Sea, have shown that the interplay of eustatic and isostatic processes continued until a few thousand years ago (Mitchell and Stephens 1974; Stephens and McCabe 1977; Synge 1977).

The rate of ice withdrawal and downwasting in the period following the Drumlin Readvance is not well known in northeastern Ireland, but the extensive areas of drumlins, and their associated tills, suggests that there was little active ice movement. The drumlin forms are perfectly developed, with little or no disturbance of the upper till which is involved in their construction over large areas of Co. Down, Co. Armagh, and parts of the Lower Bann valley in Co. Antrim. Esker and kame systems, together with limited morainic ridges, near Kilrea, Portglenone, Bellaghy and Tobermore in the Bann valley, and between Lisburn and Belfast in the Lagan valley, suggest halt stages during the period of ice wasting and retreat, but these cannot yet be dated accurately (Creighton 1974). Late-glacial sediments have been recovered from sections in old brick pits in south Belfast, where extensive deposits of laminated sands, silts and clays are exposed. The final stages of ice-wasting in eastern Ulster probably involved wholesale stagnation, downwasting, and withdrawal inwards towards the Lough Neagh lowland, allowing hill masses such as the Antrim Plateau, and Slieve Croob to become progressively free of encircling ice. Similarly in southern Co. Tyrone the ice downwasted to expose fully the Sperrin Mountains, Slieve Gallion, and isolated hill masses overlooking the Lough Erne lowland and the Clogher valley. The widespread distribution of ice-wedge casts in sand and gravel deposits, and extensive screes and patterned ground features in areas formerly occupied by Drumlin Readvance stage ice, all testify to the severity of the late-glacial climate (Zones I to III; Colhoun 1972; Stephens and Synge 1965; Stephens *et al.* 1975).

The Central Lowland

The most noteworthy glacial deposits mantling the low-lying Carboniferous limestone plain are Midlandian in age. There are extensive areas of limestone till, morainic sands and gravels, some clusters of drumlins, but especially striking are numerous long esker chains. Two fine examples have been chosen for illustration in fig. 6.26, near Tullamore, Co. Offaly, and near Trim, Co. Meath (Synge 1950, 1970).

The esker systems lie for the most part within the outer limit of the Drumlin Readvance Moraine (stage C, fig. 6.2, p.121), or are associated with the Galtrim Moraine system near Trim (stage B, fig. 6.2). The eskers are believed to have formed as ice dissolution took place in a westerly or northwesterly direction, the subglacial drainage system flowing eastwards or southeastwards. The precise manner of ice dissolution in Co. Mayo and Co. Galway is not known, nor is the direction of drainage associated with the Tuam-Swinford esker system.

The eskers around Tullamore are among the finest in Ireland. Long, sinuous ridges, alternately narrowing and widening, extend across the otherwise rather featureless country of the Central

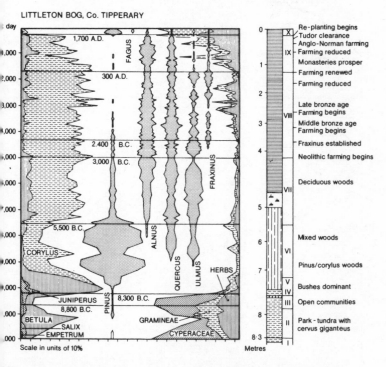

Fig. 6.25 Littleton Bog, Co. Tipperary. A composite diagram to show the pollen record from Late- to Post-Glacial time, covering a period from approximately 12 000 years B.P. to the present day. Redrawn from a diagram by Mitchell.

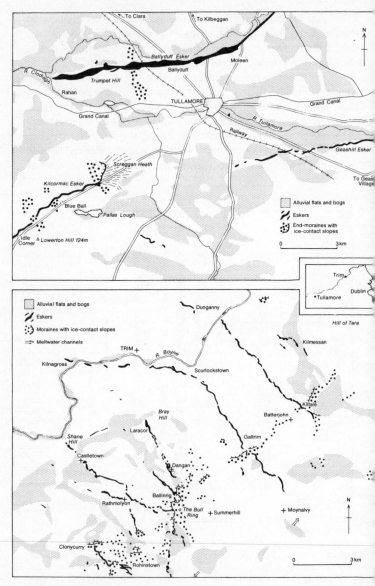

Fig. 6.26   Esker systems near Tullamore, Co. Offaly, and Trim, Co. Meath, redrawn from maps by Farrington and Synge.

Lowland, rising 10 to 15 m above both the dark peat bogs and the cultivated fields. Cross-sections of the esker ridges show considerable variation, from narrow, sharp-crested features a few metres wide, to broad, gently convex forms. All types, however, tend to have steep sides, approximating to the angle of rest of the sands and gravels forming the ridge. Transverse arched bedding and contorted ice-push structures are sometimes seen in sections in the esker ridges, while in places end-moraines interrupt the continuity of the ridges (fig. 6.26). At Trim the esker systems appear to have led subglacial drainage towards an impounded proglacial lake system associated with the line of a series of ice-marginal deltas making up part of the Galtrim Moraine. In both areas extensive sand and gravel workings continually destroy old sections in the esker systems and expose new sections.

The three-fold sequence of late-glacial deposits shown in fig. 6.23 (p.175) is typical of the sequence recorded at a number of sites, such as Cannon's Lough (Smith 1961) and Roddan's Port (Morrison and Stephens 1965), as well as elsewhere throughout the island. Smith (1970) has reviewed the evidence obtained from the pollen analysis of such a sedimentary series, while Watts (1963) and Singh (1970) have provided additional important data. The differences between the vegetational and climatic conditions of Pollen Zones I-II-III have been indicated as matching broadly those recorded in Denmark. It has been indicated that the relatively abundant sward of shrubby and herbaceous tundra plants present in the Allerød, or Pollen Zone II, supported herds of the great Irish Deer or Irish Elk, reindeer, and other herbivorous animals (Mitchell and Parkes 1949; Mitchell 1941). The transition from late- to postglacial climatic conditions is recorded in pollen diagrams such as that reproduced in fig. 6.24, compiled from investigations by Watts (1963) from a site in Co. Kerry and Smith (1961) in the north of Ireland. Remarkable vegetational changes occurred as the first closed forests replaced the former tundra vegetation, and solifluction activity all but ceased.

Mitchell (1965) has produced a pollen diagram from Littleton Bog, Co. Tipperary, which indicates the probable vegetational development for much of lowland Ireland from the late-glacial period to the present (fig. 6.25). Smith (1970) has provided us with more conventional pollen diagrams and a careful analysis of the many remaining problems of the vegetational and climatic history of Ireland. The latter part of the postglacial period has also seen the appearance of man in Ireland, with consequent vegetational modifications. It is at this point that the palaeo-ecologist and archaeologist take over the story of the subsequent evolution of the landscape (Mitchell 1976).

7 Coastline

Even if we exclude all its minor intricacies, the Irish coastline still has a length of over 3200 km, and within that distance it displays an enormous variety of geomorphic feature. The general physiography of the coastal lands is displayed in fig. 7.1, where the contribution of the dominant structural units is emphasised, as, for example, in the case of the Armorican folds in the southwest and the Tertiary plateau basalts in the northeast. On the smaller scale, much of the detail of the coastline is a result of Quaternary fluctuations of sea-level. The position of the earlier Quaternary shorelines — both submerged and elevated — is imperfectly known, but better understood are the shorelines resulting from the complex series of eustatic and isostatic movements that occurred during the later Pleistocene. In some places marine deposits are interbedded with glacial drifts, while in other places the marine deposits transgress the latest (Midlandian) tills. Organic horizons associated with the marine deposits, coupled with archaeological finds, have in a few localities provided an accurate dating of the sea's transgression and regression, and it has now proved possible to establish a chronological table for coastal events of late Pleistocene and postglacial age (fig. 7.2).

There can be no doubt of the importance of these relative changes of sea-level in accounting for many geomorphic features, but the diversity of the coastal morphology owes much to processes which have operated for less than 5000 years — the time span during which the sea has maintained approximately its present level against the island. Waves; currents; the changing supply of sediment; the fluctuating strength of the climatic elements; the activities of man — all these have to varying degrees affected the shape of the island.

Fig. 7.1   Some important structural elements and morphological features of the coastline of Ireland.

Considerable differences in the rate of coastal change are of course to be expected in the contrasted environments of the Atlantic and Irish Sea coasts; the first dominated by ocean swell, the second by wind-driven waves of variable fetch resulting from the swift passage of frequent frontal depressions. Some present-day processes are suggested in fig. 7.3, although considerable subjective judgement has been necessary to supplement the information from those few sites at which quantitative data is available. Comparison of figs. 7.1, 7.2, and 7.3 does, nevertheless, provide some insight into the complexities of Ireland's coastal geomorphology, and the three figures are basic to the province-by-province discussion of the coastline that follows.

Leinster: Counties Louth, Meath, Dublin, Wicklow and Wexford

In the physiographic sense this is the least spectacular section of Ireland's coastline. The coastline cuts across the regional strike of the Palaeozoic rocks, and Howth Head, Dalkey Head, Bray Head, Wicklow Head and Arklow Head are the most prominent of the relatively few rocky projections breaking the otherwise smooth outline of the Irish Sea coastline. Rock-cliffs seldom exceed 30 m in height on these headlands, and while abrasion platforms and low rock-reefs are common, by far the greater part of the coastline consists of low cliffs cut in glacial drifts. Between Carlingford Lough and Dublin Bay these cliffs are frequently cut in hummocky drift, which accounts for some variation in cliff morphology and height. Fifteen metres would perhaps be a maximum cliff height along this northern section of the coast, and even that is found only where active erosion has resulted in sheer cliff profiles as, for example, near Rathcor [J183049] in the Carlingford peninsula.

Recent constructional forms are conspicuous, these including the sand-spits and dune systems between Howth and Rush, the fine tombolo joining Howth to the mainland, the shingle and sand-bars at Termonfeckin [0155809], and the extensive intertidal flats in Dundalk Bay. There are also stretches of 'dead' cliffs to the north of Benhead [0176687] and between Dundalk and Rathcor, where progradation by shingle-bars and sand-dunes has occurred. Elevated shoreline notches and raised shingle-bars of late-glacial and post-glacial age have been mapped between Dundalk Bay and Dublin.

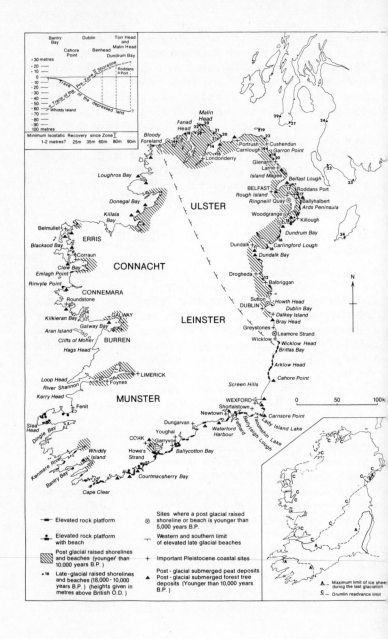

The late-glacial beaches decline in height southwards, being well exposed at White's Town [J232066] in the Carlingford peninsula, and at Port [0150891], Co. Louth. The beach deposits have been cryoturbated by frost action, contain no shell fauna, and truncate the surfaces of the glacial drifts. It has been shown that these beaches are contemporaneous with certain morainic systems which reach the present coastline (McCabe 1973), the strongest morainic line being that of the Drumlin Readvance (Stephens and Synge 1966a; Stephens *et al*. 1975). Late-glacial beaches occur only outside the Drumlin Readvance morainic system in Co. Louth and Co. Down, and, from evidence obtained elsewhere, the beach has been dated at approximately 18 000 years B.P. (Stephens and McCabe 1977).

The postglacial raised-beach extends southwards to Dublin Bay, where it is clearly represented both in the tombolo that joins Howth to the mainland, and in the sedimentary series within Dublin Bay (Naylor 1965). The tombolo stands at 4 m above mean sea-level, and at Sutton the shingle rests upon a kitchen-midden containing charcoal which has a 14C age of 5250 ± 110 years B.P. The maximum of the transgression is clearly younger; further evidence at Sutton and from Dalkey Island, indicates regression by stages from the 4-m line. No elevated notches or remnants of a beach have been identified to the south of Wicklow Head, but at Leamore [0315053] a barrier-beach-bar may have been breached about 5000 years B.P., resulting in a marine incursion across the site of the present coastal marshes to a height of 3.75 m above mean sea-level. The evidence and the implications of the various 14C dates have been examined from several sites on the east coast of Ireland (Mitchell and Stephens 1974). It seems possible that a marine transgression to 3 or 4 m above mean sea-level did occur about 5000 years B.P., and that as a consequence some of the elevated postglacial notches and beach systems are somewhat younger — perhaps as much as 2000 years younger — than previously thought possible (Stephens and Synge 1966a; Stephens 1968).

Remnants of an old wave-planated and ice-striated rock surface

---

Fig. 7.2   Some important Pleistocene and Holocene features of the coastline of Ireland. The inset maps show the approximate extent of the Midlandian ice-sheets and the degree of isostatic recovery which has affected the late-glacial (pre-Pollen Zone II) shoreline.

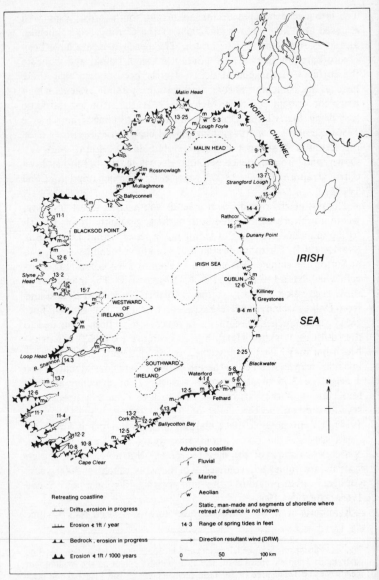

Fig. 7.3 Retreating and advancing segments of the coastline, together with the range of spring tides and the Direction Resultant Wind direction (DRW, after the method of A. Schou).

are exposed intermittently to the south of Wicklow Head, notably at Clogga Head [T258691], Kilmichael Point [T257664] and Courtown [T202561], but elsewhere the surface is completely hidden from view by the suites of Pleistocene deposits found on this part of the coast, or, as in Brittas Bay [T310830], by modern sand-dune accumulations. No ancient elevated beach-deposits have been recorded, and although there are some examples of 'dead' cliffs at Brittas Bay and near Arklow, the available evidence indicates modern progradation of the shoreline rather than elevation.

Between Cahore Point and Wexford town there are unstable cliffs up to 25 m high in glacial drifts, the greatest heights occurring near Blackwater where the Screen Hills morainic.complex is intersected (Stephens 1970). Even here there are alternating segments of 'live' and 'dead' cliffs, the latter formed where a narrow zone of temporary foredunes has been partially stabilised by marram and other dune grasses.

At Wexford town, where the Slaney reaches the sea, a large coastal indentation has been partially reclaimed within the shelter of converging spits. Still farther south, at Ireland's southeastern corner, granite forms the low headland of Carnsore Point where a thin drift cover overlies both ice-abraded and wave-abraded surfaces of well-jointed rock. Copious quantities of weathered and eroded joint-blocks litter the foreshore, and are found as far west as Kilmore Quay as a result of ice-carriage.

The south Wexford coast consists of low, rocky headlands and broad shallow bays, where fine sand-spits enclose a sequence of lagoons — Lady's Island Lake, Tacumshin Lake and Ballyteige Lough (fig. 7.4). Important changes have occurred in these spits even since the first editions of the Ordnance Survey 1:10560 maps were published in the 1840s, and it is known that submerged peat and forest-beds occur at a number of sites (see fig. 7.2, p.184). Where Pleistocene deposits constitute the cliffline they usually overlie an abrasion surface together with ancient beach deposits and spectacular slumping of the unstable materials has permitted considerable retreat, as, for example, at Fethard [S791049] (cliffs in head and till) and at St Helen's [T145099] (till). At Ballyhealy [T012047] temporary sand-dune formations provide some protection to low cliffs (less than 3 m high).

The broad and deeply drowned valley of Waterford Harbour constitutes a formidable morphological feature in direct contrast to the spit-enclosed bays to the east. A buried channel has been detected, and the enclosing cliffs can reach 15 m in height. They are often sheer and cut in rock or thick glacial drifts, and they may be subject to severe

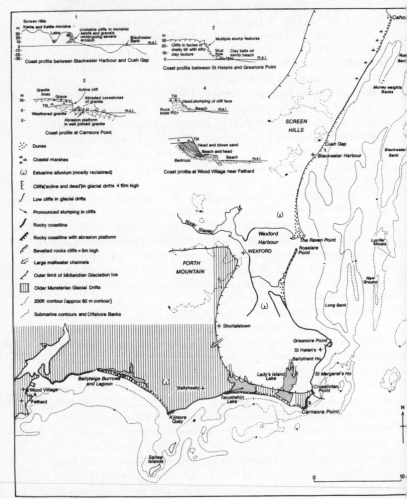

Fig. 7.4 Geomorphic features of the coastline of southeastern Co. Wexford. Five representative cross-sections serve to illustrate the varied morphology of a coastline cut mainly in superficial deposits.

scouring. An important Pleistocene site is known at Newtown [S704067] in the western cliffs of the estuary (Mitchell 1962; Watts 1970).

Munster: Counties Waterford, Cork, Kerry, Limerick and Clare

Westward of Waterford Harbour the degree of control exerted by the regional geological strike increases (see fig. 7.1, p.182), although the deep-water rias of Cork Harbour, and, to a lesser extent, Kinsale Harbour and Clonakilty Bay, do cut across the geological and topographical grain. The old valley systems sometimes have considerable alluvial infilling (e.g., Dungarvan Harbour which also possesses a fine enclosing bay-bar), and these sediments mask buried rock-channels extending to over 30 m below datum. Prominent cliffs and headlands, such as Helvick Head, Knockadoon Head, Power Head and Old Head of Kinsale, are usually developed on the more resistant Old Red sandstone or Carboniferous sandstone. The cliff profiles are generally bevelled and then topped by portions of the very striking planation surfaces that were discussed in chapter 4. Cliffs in rock become more conspicuous and higher as one proceeds westwards, and it is in Ballycotton Bay [X010660] that we can see the last really extensive cliff development in Pleistocene deposits, although a very important site occurs also at Howe's Strand [W559425] in Courtmacsherry Bay.

At Garryvoe and Shanagarry, in Ballycotton Bay, at least 200 metres of drift have been eroded in the last 120 years, and sections are exposed in descending sequences of till, outwash sands, periglacial head and beach deposits, all overlying an old rock abrasion-platform. Similar sequences are known at many points on the southern coast (Mitchell 1962; Synge 1963a, 1964; Farrington 1966b). The 'preglacial' cliff-notch is usually hidden. It now seems likely that more than one such notch is buried below the coastal drifts. A lower shoreline, a few metres above mean sea-level, is represented by beach-gravels and is probably Last Interglacial, or Ipswichian, in age, and a higher shoreline, perhaps 20-5 m above mean sea-level, is indicated by rather indistinct benches and notches masked by till and head. The higher shoreline is probably of Gortian (= Hoxnian) age, but so far no good beach-deposits have been detected. A revision of the probable age of the lower, buried beach in counties Waterford and Cork stems not from a reinterpretation of the age of the superincumbent Munsterian Irish Sea or Ballycroneen shelly till (Mitchell 1972), but from other evidence obtained recently by officers of the Geological Survey of Ireland, and principally by Warren and Synge. It is now believed that on the east coast, where the abrasion-platform is striated (at Courtown Harbour, for example), the overlying till is of Midlandian age. No beach-deposits are known from the east coast sections, whereas at many sites on the south coast, the abrasion-platform is not striated (Garryvoe, for example, and Shanagarry), and the buried beach-gravels rest upon a wave-pounded surface. The overlying tills and periglacial deposits are believed to have slumped on a large scale into their present positions, and hence 'older' drifts are resting upon a Last Inter-glacial (Ipswichian) beach and shore platform. In places, elements of the rock-platform may of course be older than Ipswichian, but Synge suggests that the notch representing the cliffline of this beach can be seen to pass off the rock and into the 'older' drift, as, for example, at Whiting Bay, Co. Waterford. Thus, one of the most difficult problems concerning the age of the buried beach-gravels and their copious erratics, may at last be nearer to solution. It is not claimed that this explanation fits every site, but it is certainly supported by the work of Bowen (1973) in South Wales, and of Kidson and Wood (1974) in southwestern England. Certain vital sections, such as that at Newtown in Waterford Harbour (see p.139) now require urgent re-examination. If the interglacial deposits resting upon the shore-platform, and in association with beach-gravels, should prove to be buried only by slumped Munsterian till, then further support is provided for Synge's hypothesis. Undoubt-edly, this problem will be the subject of further field-work, and not a little controversy in the years to come.

At Cape Clear there is a sharp change in the orientation of the coastline, a change also reflected in the trend of the submarine contours. Northwest to southeast trending faults may have deter-mined the regional coastal trend before the etching out of the large inlets of Dingle Bay, Kenmare River, Bantry Bay, Dunmanus Bay and Roaringwater Bay (fig. 7.5). The blunted headlands of the southwestern peninsulas, and the multiplicity of islands and sea-stacks, testify to the power of the Atlantic waves. Large quantities of glacial drift have been removed from the lower part of the coastal slope on many segments of the peninsulas, and there fragments of

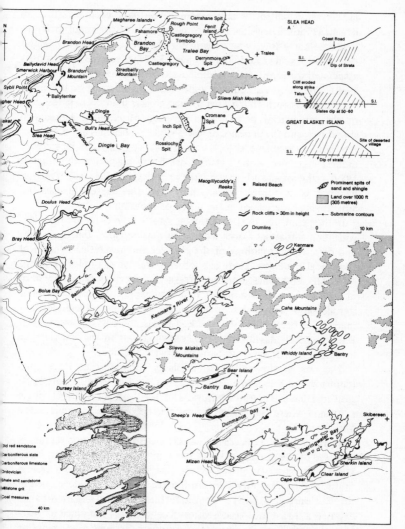

Fig. 7.5 Geomorphic features of the coastline of the southwestern peninsulas. Three cliff profiles are illustrated (A, B, C) to show the varied morphology at the western extremity of the Dingle Peninsula (long, head-mantled, sub-aerial and convex slopes leading to the marine cliffs), where hog-backed cliff forms are found.

old abrasion platforms and beach-gravels have been recorded by Bryant (1966). There has been some erosion of the small drumlin-field in Bantry Bay, where deep borings made on Whiddy Island [V960500] revealed Zone II (14C age 11-12 000 years B.P.) deposits associated with fresh-water sediments at 57 m below datum (Stillman 1968). The site is particularly important in helping to establish the position of a southern Irish sea-level at greater than -50 m in this part of the late-glacial period. The absence of elevated late-glacial shorelines, and the presence of low postglacial beaches at only a very few points (Stephens 1970; Lewis 1974), suggest that in spite of modest glaciation — the Lesser Cork-Kerry glaciation (Farrington 1959) — the area can be regarded as relatively stable. Apparently it has not been subject to anything like the degree of isostatic adjustment experienced in the north of the island.

In the peninsulas of the southwest most of the cliffs are bevelled, with long sub-aerial slopes leading to marine cliffs, but the height and detailed morphology of the cliffs commonly reflects the local geological structure. There are some impressive examples of recent rock-falls, and some apparently fresh vertical cliffs exceeding 30 m in height where the sea has been able to exploit the fissile Carboniferous slates and shales. In the Dingle peninsula wave-attack has made large breaches in the Old Red sandstone cliffs between Sybil Point and Ballydavid Head, thus allowing part of the Ballyferriter lowland to be drowned to form Smerwick Harbour.

The rias of the southwest in places contain some large spits (Guilcher and King 1961; Guilcher 1965) and fine examples are to be seen in Dingle Bay and northward of Castlegregory in the Dingle peninsula where the Castlegregory spit lies between Brandon Bay and Tralee Bay.

To the north of Tralee Bay, the entrance to the Shannon estuary is enclosed by the prominent rocky headlands of Kerry Head and Loop Head. The estuary itself follows the line of a series of fold axes and faults, and its shallow waters are flanked by low cliffs developed in till. Widespread silting is evidenced by the large sand and mud-banks exposed at low-water, and Scattery Island marks the place between Kilrush and Tarbert where the Southern Ireland End Moraine (Midlandian Glaciation) crosses the estuary (see fig. 6.2, p.121).

The northern shores of the estuary lie in Co. Clare, and the Atlantic coast of that county, from Loop Head to Liscannor Bay, displays bevelled cliffs developed in Namurian shales and carrying

pockets of glacial drift and solifluction debris. Immediate...
north of Liscannor Bay lies one of Ireland's most famed piece...
coastal scenery — the Cliffs of Moher. The cliffs extend for 8 km,
are 122 to 196 m high, and occur where the Namurian grits, flagstones,
and shales of the Clare Plateau are undergoing vigorous marine erosion. The cliff-profiles are usually vertical or stepped, and a number of
spectacular sea-stacks reveal the lithological divisions and joint planes
which have determined the direction of wave-attack. Two final points
about the cliffs are worthy of note. Firstly, there is no evidence
of the presence of wave-cut platforms at the foot of the cliffs, and
secondly, at the top of the cliffs the regional slope is not towards the
Atlantic, but rather inland towards the valley of the Dealagh river.

Northwestern Clare is the land of the Burren limestones, and from
Fisherstreet northwards the coastline offers a superb display of tier
upon tier of limestone pavements. The area around Black Head must
be a paradise for the student of pavement-form although around
Murroogh, to the south of the head, the limestone disappears
temporarily beneath wind-blown sand and low mounds of glacial
drift. In other places there are also storm-beaches banked against the
limestone outcrops, and at the coast itself there are a number of
elevated caves exposed to the sea. Although the Aran Islands form a
part of Co. Galway, their coastline is most conveniently mentioned
here because in many respects it reproduces the morphology of the
Burren coastline. There are the same low coastal cliffs, the same
patches of blown-sand and glacial drift, the same terraced pavements, and the same strange solution features on the shore where the
limestone has in places been reduced to something which at first
sight looks remarkably like a scoriaceous lava. In the islands the
most spectacular coastal scenery is on the southwestern coast of
Inishmore where the limestones dip towards the sea and form a series
of vertical or overhanging cliffs up to 90 m high. In many places large
block-falls have occurred, while near the southeastern corner of the
island vertical joints have been opened out to form some interesting
blow-holes. On the mainland to the east of Black Head the limestones continue to display their strongly terraced character where the
northern face of the Burren Plateau overlooks the waters of Galway
Bay. Here, however, the limestone near the shore is widely obscured
by spreads of glacial drift and blown-sand, and among these drifts
there are a number of hollows which contain lakes whenever a high
tide causes a back-up in the ground-water circulation.

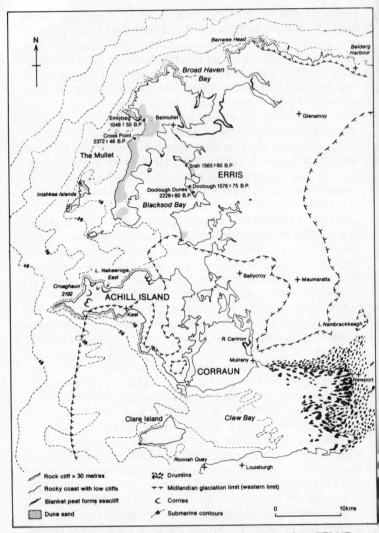

Fig. 7.6 Some geomorphic features of the coast of Clew Bay and north-western Co. Mayo. Extensive areas of blanket peat in Broad Haven and Blacksod Bay pass without break below high-water mark, and are cliffed. The dune systems on The Mullet contain buried peat and relic soil horizons. This map is based upon the work of Charlesworth, Farrington, and Synge. R.W. Crofts has allowed the 14C dates from buried organic horizons in the

Connacht: Counties Galway, Mayo, Sligo and Leitrim

The Central Lowland reaches the sea at the head of Galway Bay, and the subdued but intricate coastal morphology around Kinvarra and Clarinbridge is related chiefly to the discontinuous mantle of drift that overlies the limestone. Very different is the northern shore of Galway Bay to the west of Galway town (see fig. 3.7, p.47). There the coastline is developed in the Galway granite, and its remarkably straight form between Galway and Inveran is perhaps a result of the tilting and partial submergence of an ancient planation surface bevelled across the granite body (see p.48). West of Inveran the coastal configuration changes dramatically, and between Cashla Bay and Roundstone the granite forms an island-studded, and deeply indented coastline, where jointing contributes some degree of control to the minor feature of coastal morphology. Along the coast as a whole wave-abrasion is restricted, and in most places the inter-tidal zone consists of granite bedrock, granite blocks, and small pocket-beaches of sand and shingle, backed in places by low fringing-dunes. Very limited exposures of low, postglacial beaches have been recorded at the head of Galway Bay by Farrington (1964).

If we overlook the many intricacies of the drowned coastline of the Iar-Connacht Lowland and the Killary Mountains, then we are left with a coastline which in its broad outline displays some very abrupt changes of direction. At Gorumna Island and Golam Head the east-to-west trend of the northern shore of Galway Bay is replaced by a northwesterly trend as far as Slyne Head, then by a northerly trend to Aughrus More, by an east-north-easterly trend to the mouth of Killary Harbour, and thence by a northerly trend to Roonah Quay. In some places, at least, this pattern is reflected in the changing trend of the steepened submarine slope about 16 km offshore, and some kind of structural control is therefore probably responsible. To the north of the Killary Mountains, Clew Bay certainly coincides with a down-warped and down-faulted trough (fig. 7.6).

---

coastal dunes to be quoted from his unpublished investigations on these dune systems. Although the 14C dates must be regarded with some reservations (several periods of dune stability are indicated) these do not appear necessarily to match those so far recorded at Dundrum, Co. Down, on the Irish Sea coast.

The coastline of Clew Bay offers a marked contrast to the coastline of the regions both to the north and south of the inlet. To the south of the bay the rocky coastline of western Co. Galway displays many deep indentations, of which the fiord-like Killary Harbour is much the most impressive, and there are some fine abrasion platforms emerging from beneath the drift cover. At Rinvyle Point [L644642], for instance, up to 400 m of shore-platform is backed by low cliffs formed in multiple till units. Farther to the north, in southwestern Mayo, the coastline south of Emlagh Point is developed in a low, featureless plain where cliffs rarely exceed 6 m in height and are actively retreating. Some fine storm-shingle-ridges are piled against the till cliffs, and in places rest upon either blanket or basin-peats lying in the inter-tidal zone.

Inside Clew Bay the coastline is formed chiefly in till where a cluster of east-to-west trending drumlins has been drowned to form a remarkable archipelago. The Drumlin Readvance ice-limit extended from beyond Clare Island to Emlagh Point and Roonah Quay, and thence across the bay to the Corraun peninsula, but many of the most westerly of the drumlins have been removed by the sea. Others have west-facing cliffs 10 to 30 m high blunting the exposed ends of the drumlins, and between the drumlins a series of complex spits and tombolos has been constructed (Guilcher 1962a). One of the rewards for an ascent of Croagh Patrick on the southern shores of the bay is a magnificent prospect over the dozens of drumlin-islands lying to the north; tradition maintains that they number 365.

At the western end of Clew Bay, Clare Island and Achill Island contain some of the finest cliff scenery to be found in Ireland. The vertical element in the cliffs can exceed 300 m with scree mantling the bevelled sub-aerial slopes above the vertical cliffs. In Achill one fascinating feature of the high western cliffs around Croaghaun is a group of cirques perched at various levels above the sea (see p.52).

Blacksod Bay extends northwards from Achill Island and becomes enclosed by the Mullet peninsula, while northwards again Broad Haven forms a major inlet, also abounding in creeks, headlands and small islands (see fig. 7.6, p.194). The Mullet is composed of a series of low rocky islands and reefs of granite and metamorphic rocks, joined together by plugs of glacial drift, as well as by spits and tombolos. Mobile sand-dunes have over-run the Atlantic side of the peninsula and remain unstable. The dune systems are composite, containing in places dark bands which represent parts of the soil horizons developed on former fixed dunes which have subsequently been buried by moving sand. The radiocarbon dates obtained so far from the buried organic horizons suggest that there have been two, or possibly three periods of dune stability at about 1000, 1500 and 2300 years B.P. Dune construction and generally mobile sand conditions existed in the intervening periods, although rates of sand accumulation are unknown (Crofts 1971).

In both Broad Haven and Blacksod Bay the sea has cut cliffs 1-2 m high in blanket-peat, which extends below low-water mark, and pine stumps can be seen in a position of growth amidst the peat. The general absence of elevated beaches or shorelines is noteworthy, although there are a few indications of terraces or notches, and some shingle-ridges, all at a low-level, and never more than 1-3 m above mean sea-level. While these features resemble those seen notching the drumlins in Bantry Bay, along the northern shore of Galway Bay, and at a few places in Clew Bay, their age is still in doubt. Some terraces may not be beaches or shorelines at all, while further investigation is necessary to consider the possibility that some of the low (1-3 m) shingle-ridges are modern, and related only to exceptional storm-wave surges. Anyone who has worked on the exposed Atlantic coast of Ireland will know of the considerable variation in height which can be expected over short distances for wave-constructed features.

The northern coast of Co. Mayo has a general trend from slightly north of west to slightly south of east, beginning with the schists and gneisses of Benwee Head, and passing thence onto the Carboniferous strata of the Ballina syncline. The nearness of deep water to the coast west of Killala Bay again seems to emphasise the possibility that the general pattern of the coast is a result of structural control. Blacksod Bay is certainly aligned along a north-to-south trending fault-system, and similarly orientated structures perhaps account for the sudden steepening of the submarine relief a few kilometres off the Mullet peninsula. Between Benwee Head and Belderg Harbour nearly vertical cliffs in schist and gneiss may reach heights of 260 m, their scenic quality enhanced by the topographical slope being towards the south and thus away from the cliff-edge. The coastal slope carried a regolith made up of a mixture of glacial drift and head deposits, some of the latter being derived from former land areas lying to the north of the present cliffs. There are some spectacular sea-stacks such as those around Benwee Head at the four Stags of Broad Haven, Doonvinalla and Pig Island, while deep, structurally guided geos form spectacular rock-bound inlets such as that at Porturlin [F888428].

Farther to the east, towards Downpatrick Head, flat-topped cliffs

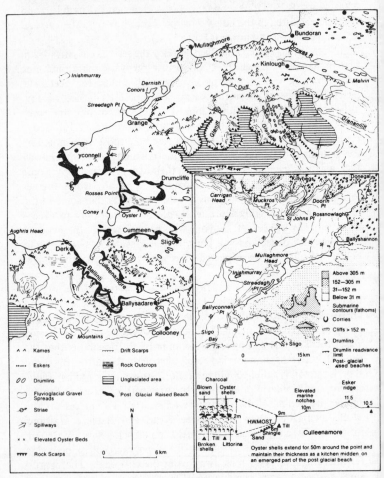

Fig. 7.7   Some geomorphic features on the coastline around Sligo Bay and Donegal Bay. A sketch shows one of the many elevated oyster beds found in Sligo Bay and associated with the post-glacial raised beach. (With acknowledgement to F. M. Synge.)

developed in gently dipping Carboniferous strata exceed 80 m in height, but they are still vertical cliffs, sometimes with pronounced overhangs and with caverns at their base (e.g., near Benaderreen). Killala Bay is unusually wide and rectangular in plan, which suggests structural control, but high rock-cliffs are generally absent, and are replaced by abrasion platforms and low cliffs in glacial drift. The estuary of the River Moy is partially blocked at its seaward end by spits and by Bartragh Island, where rock-outcrops have been linked by marine deposition. The rocky coast continues eastward to Aughris Head, but cliffs nowhere exceed 40 m and are generally much lower. Once Sligo Bay is reached rock-outcrops become less important, and a variety of drift forms — eskers, kames and drumlins — are found, together with some fine spits carrying active dune systems (fig. 7.7). At Derkmore Point [G575360] we find the first exposure of a postglacial raised-beach, a feature which can be seen as a distinct element a few metres above mean sea-level around much of Sligo Bay and Ballysadare Bay. Associated with the raised-beach are a number of oyster-shell beds, some of which are undoubtedly kitchen-middens, such as that at Culleenamore [G610341] (fig. 7.7). At a few sites, such as near Oyster Island (G630394], marine sediments interdigitate with freshwater marls, and there have probably been changes in the position of the shoreline as drift barriers have been breached allowing freshwater lakes to be transgressed, without necessarily any change of level between land and sea.

North of Sligo Bay and Drumcliff Bay the coastal plain is about 8-10 km wide, ending sharply at the foot of the Dartry Plateau Country (fig. 7.7). For much of its width the limestone bedrock is masked by kames and morainic deposits, but there are some coastal exposures, for example on the promontory of Mullaghmore. Here wave-action can be observed 'quarrying' the well-jointed sandstone, and a variety of marine solution forms can be examined on the abrasion platforms (fig. 7.8). The postglacial raised-beach is conspicuous by its absence from Ballyconnell northwards until one reaches the drumlin archipelago at the head of Donegal Bay, but there is a fine spit linking Streedagh Point and Conors Island [G650510].

Ulster: Counties Donegal, Londonderry, Antrim and Down

To the north of the Erne estuary irregular outcrops of Carboniferous limestone and shale form low headlands at the seaward edge of the drift-masked coastal-plain. But the coastal topography soon becomes dominated by northeast to southwest trending drumlins, many of which form islands, or are linked together and to the mainland by a variety of spits, bars and tombolos. The postglacial raised-beach re-appears to the north of Rossnowlagh Lower [G856690], and has been traced, standing at between 1 to 3 m above mean sea-level, among the drumlins at the head of Donegal Bay and in Inver Bay,

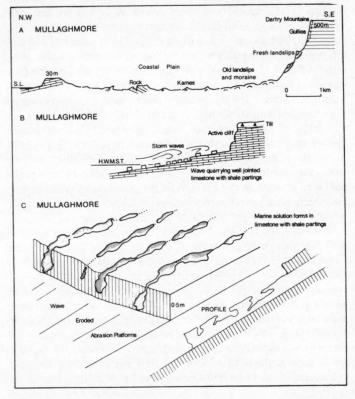

Fig. 7.8 A cross-section from the Carboniferous limestone plateau of the Dartry Mountains to Mullaghmore. Glacial drifts obscure much of the coastal plain which terminates at the foot of the limestone cliffs where both old and fresh landslides of the rotational type are found. Wave erosion and marine solution forms are particularly well developed near Mullaghmore.

but it does not appear to be present to the west of the St John's peninsula. Perhaps there it may be low enough to have been destroyed or overlain by modern storm-beaches (see fig. 7.7, p.198).

The strongly impressed Caledonian structures of Donegal are well reflected in the county's coastline, but the presence within the county of Tertiary dyke systems trending from northwest to southeast suggests that some, at least, of the similarly orientated coastal inlets may be related to a more recent fracture-pattern which has been imposed upon the dominant Caledonian strike. To the northwest of Malin Head, and between the northern coast of Inishowen and Inishtrahull, there is a marked depression on the sea-floor which may reflect the presence of northwestward striking faults. Similarly a major Tertiary fault-zone may explain the general coastal trend between Malin Head and Inishowen Head — a fault-zone which would intersect the Lough Foyle Caledonian fault-system beyond the mouth of Lough Foyle.

Doorin Point, St John's peninsula, Drumanoo Head, and Muckros Head all reflect the dominant northeast to southwest geological strike, but it is not until Carrigan Head is passed that the finest and highest cliffs are reached. Below the summit of Slieve League (598 m) scree-clad slopes dip steeply at over 30° before eventually giving way to nearly vertical cliffs 100 to 200 m high. In some places massive quartzite tors or buttresses 6-12 m high stand out above the scree, and probably indicate the severe periglacial denudation of these coastal slopes during the Midlandian glaciation (fig. 7.9).

Patches of Older Drifts, including shelly till of Scottish origin, are exposed on the much lower coastal slopes and cliff top near Malin Beg (G490800] where there is occurring severe coast erosion of both rock and superficial deposits. But cliffs exceeding a height of 180 m appear on the metamorphic rocks and quartzites below Slievetooey, on the southern side of Loughros Bay. At the head of this bay large drumlins appear around Ardara, and in some places there is a suggestion of a low postglacial marine-terrace partly obscured by sand-dunes. Organic horizons have been observed at a number of points in inter-tidal situations among the drumlins, but as yet these and other known sites in Co. Donegal have not been examined in detail.

To the north of Loughros Bay forms resulting from glacial erosion become more conspicuous, especially on the ice-scoured slopes of the Rosses Granite and in the island-strewn areas of Sheep Haven,

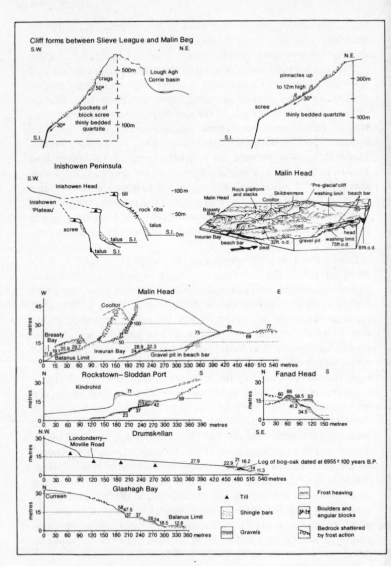

Fig. 7.9 Some geomorphic features of cliffs in Co. Donegal, redrawn from diagrams by Stephens and Synge.

Mulroy Bay and Lough Swilly. Vertical marine-cliffs exceeding 150 m in height are found on the quartzites and dolerites that form the prominent headland of Horn Head [C012420], and also on the exposed west-facing cliffs of Aran Island, but the cliffs are somewhat lower on Bloody Foreland [B820346]. Gweebarra Bay and other major indentations of the coast are in many places obstructed by bay-bars and spits of shingle and sand, these sometimes forming a base for unstabilised dune systems. Massive storm-beaches of shingle and cobbles occur where Atlantic waves have had access to the glacial drifts, or where bedrock has yielded to prolonged pounding, some of the finest examples being seen south of Bloody Foreland, and at Fanad Head, Rockstown [C330485] and Malin Head. Constant pounding by wave-action has produced vast quantities of quartz sand, supplemented in places by shell sand, and consequently the Donegal coast is well endowed with fine strands, such as those in the bays of Ballyness, Ballyhiernan and Pollan. But the copious quantities of sand have also proved to be a considerable menace to farmland and settlements adjacent to the coast. Blowing sand has long been a nuisance at Dunfanaghy, behind Horn Head. It is also recorded that at Rosapenna [C112374] blowing sand overwhelmed Lord Boyne's house and sixteen farms in 1784, burying the land to depths of 6 m in places. The instability of the sand-dunes in that particular case was attributed to an increase in the rabbit population following the human decimation of the local foxes. There can be no doubt that around the Horn Head peninsula there is still instability of the blown-sand, just as there is at a number of other places on the coast of Ireland.

Elevated late-glacial and postglacial shoreline notches and beaches have been mapped east of Bloody Foreland, culminating in the very fine exposures at Fanad Head and around Malin Head (Stephens and Synge 1965; fig. 7.9). The highest late-glacial shingle-bars and washing-limits in Ireland occur around Malin Head, and they are sometimes associated with an elevated abrasion platform and 'dead' cliff. This cliff is masked by old 'fossil' talus, composed in part of large angular quartzite blocks. The massive shingle-bars are well displayed at Ballyhillin [C390589], and the shingle is severely cryoturbated in a number of road-side exposures.

The postglacial beaches have been identified by their content of mollusca and by the absence of any frost cryoturbation features. These shorelines can be seen below the late-glacial beaches throughout

Inishowen, but especially in Lough Foyle, where interesting sites occur at Drumskellan [C494277] (Colhoun *et al.* 1973), Moville, and between Londonderry and Limavady. It is necessary to distinguish the reclaimed areas from the lowest of the raised-beaches. At Drumskellan the highest beach-bar stands at about 4 m above mean sea-level and rests upon an organic horizon dated to 6955 ± 100 years B.P. There may be a hiatus between the construction of the beach-bar and the accumulation of the forest-bed, but there can be no doubt that a marine transgression occurred after 7000 years B.P. Such a relatively low level for the highest postglacial beach is rather unexpected for an area which experienced considerable isostatic recovery in the late-glacial period. Comparison with sites such as Ballyhalbert [J645636], Co. Down, where beach-bars rise to 8 m above mean sea-level and transgress organic horizons dated to 8120 ± 130 B.P., suggests that there is indeed a hiatus between beach and forest-bed. It seems likely that the postglacial transgression into Lough Foyle may have been delayed as a result of the weak wave-action, in a relatively sheltered situation, being unable to produce rapid regression of the shoreline. Consequently, the highest postglacial raised-beach is in an unusually low position because of its relatively late emplacement. It is noteworthy that the highest postglacial shorelines near Malin Head are several metres higher than those found at Drumskellan.

Lough Foyle is aligned along a northeast to southwest synclinal axis, and its northwestern margin is defined by a major fault-zone against the Pre-Cambrian basement of Inishowen. Tertiary plateau-basalts form the high ground east of Lough Foyle, with impressive vertical or stepped cliffs occurring between Downhill and Castlerock. South of Downhill former marine-cliffs have become 'dead', partly as a result of isostatic emergence, and partly as a result of the growth of Magilligan Foreland. Narrow terraces cut both in the seaward face of some of the large landslips below Binevenagh and in morainic material mantling the lower slopes, effectively place an upper limit on the late-glacial shorelines at approximately 20 m (Carter 1975).

There are no real sea-cliffs within Lough Foyle, although here and there rock-outcrops project through a generally thick suite of glacial drifts. Low marine terraces of late- and postglacial age range up to 15 m above mean sea-level and dominate the coastal topography behind extensive inter-tidal flats and reclaimed land. Between

Londonderry and Limavady the 15-m late-glacial terrace is extensively pitted with kettle-holes, such as the two Loughs Enagh [C470195], and the upper 2 m of the marine-planated fluvio-glacial sediments have been cryoturbated, and show fossil ice-wedges.

Magilligan Foreland reduces the entrance to Lough Foyle to a channel less than 2 km wide. The foreland is undoubtedly of quite recent marine construction in its northern part, where a series of broad, curving sand-ridges can be seen. These ridges are overlain by sand-dunes, some of which are mobile on the outer coast, but peat and marl bands are also inter-bedded with the marine sands of the ridges. One 14C date so far obtained from one of the peat layers at about 6 m O.D. gave a date of 1535 ± 40 years B.P. The marine-deposited sediments seem to rest upon a thick basement of glacial morainic and outwash gravels which are known to exceed 50 m in thickness before rockhead is reached.

The basaltic plateau of northeastern Ireland is interrupted by the major depression of the Lower Bann valley, and as a result of this high cliffs are largely absent between Castlerock and the mouth of the River Bush. Numerous volcanic vents are intersected by the sheer and stepped cliffs along the plateau's northern coast, but fewer such features are encountered along the North Channel coast. Cliffs seldom exceed a height of 160 m except at Benbane Head, where the well-known Giant's Causeway can be seen (fig. 7.10). The ground plan of the Causeway's cliffs is remarkable in that it consists of a series of huge amphitheatres whose shape is probably joint-controlled. Extensive talus-slopes mask large portions of the sheer basaltic faces, and rock-benches occur at a variety of levels, probably reflecting the different rates of retreat of the exposed lava layers. Up to 30 m above mean sea-level there are indications, such as potholes and undercut-notches, that wave-abrasion has been responsible for certain of the rock-platforms, but few of them carry any beach material (fig. 7.11).

Faulting frequently brings chalk against basalt (e.g., on Rathlin Island and at White Rocks) and the coastline closely follows a fault zone across White Park Bay and beyond (fig. 7.10). Planation at 50 to 70 m has produced a low tableland across chalk and basalt, while in White Park Bay dune systems conceal landslips of basalt and chalk over Liassic clay. Major fault-zones exert further control on the coastal plan in Ballycastle Bay, where Carboniferous strata are exposed in almost sheer cliffs east of Ballycastle town. These cliffs become even more spectacular as Fair Head is reached, and heights of 180 m are attained where the olivine-dolerite of the Fair Head Sill acts as a cap rock. Prismatic jointing controls the shape of the dolerite masses which have broken away to form a talus of enormous

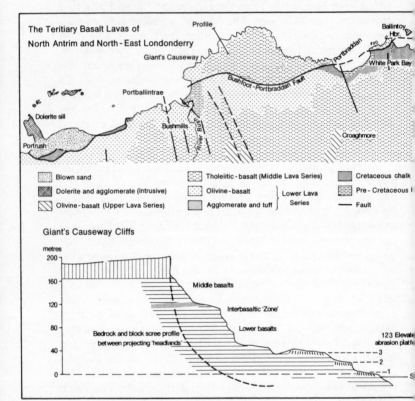

Fig. 7.10 Some geological and morphological features on the north coast of Co. Antrim between Portrush and Ballintoy Harbour. The various lava series have been faulted against one another and the underlying chalk and Lias. At Portballintrae, the mouth of the River Bush, and at White Park Bay, faulting has determined the position of indentations in the coastline. In White Park Bay exposures of Liassic shale have allowed rotational landslips to develop but these are partly covered by dune sands. The composite section at the Giant's Causeway shows the stepped cliffs, possible wave-cut platforms, and a typical profile from the columnar middle basalts and across the screes of the amphitheatres lying between the headlands.

blocks. East of Fair Head, where the dolerite sill separates into two parts, huge dolerite prisms are still falling from the free-faces, but the talus slopes also consist in part of debris derived from the Carboniferous shales (fig. 7.12).

The North Channel is believed to be aligned along northwest to

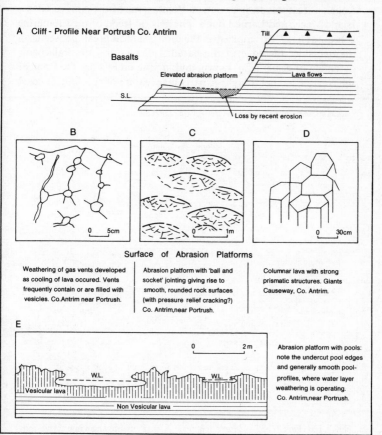

Fig. 7.11 A. This cliff-profile across olivine basalt (lower basalts) shows a well developed elevated abrasion platform (Late- and/or Post-Glacial age?) and the probable loss by recent erosion.
B, C, D, and E These are examples of the exposed surfaces of various recent abrasion platforms where different weathering systems are in progress.

southeast trending faults, and geophysical investigations have certainly established the presence of a major tectonically controlled break between northeastern Ireland and southwestern Scotland. Systems of intersecting faults control the alignment of many of the major features along this section of the coast as far south as Belfast Lough — features such as Garron Point, some of the Antrim Glens, Island Magee and Larne Lough. The general trend of the submarine contours (fig. 7.13) emphasises the topographical break between, on the one side, the edge of the basaltic plateau and the Palaeozoic basement of eastern Co. Down, and, on the other side, the floor of the North Channel trough. A particular feature of the Antrim coast

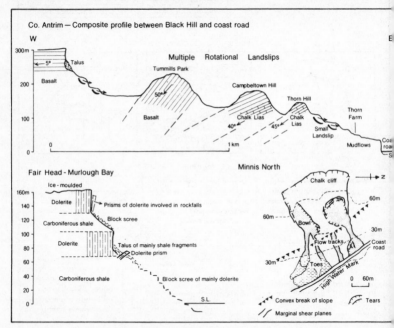

Fig. 7.12   Fair Head — Murlough Bay: A cliff profile showing active mass wasting of dolerite sills and Carboniferous shale.
Black Hill: A series of rotational landslips involving basalts, chalk, and Liassic shale, lead from the edge of the basaltic plateau to the Antrim Coast Road, between Glenarm and Larne.
Minnis North: Multiple mudflows in well-defined flow tracks affect the coastal slope of the Antrim plateau, especially between Carnlough and Larne. With acknowledgement to D.R. Archer, G.R. Douglas and D.B. Prior.

is the enormous slumped blocks with reversed slopes, involving up to 300 m of Tertiary basalts, chalk and Liassic shale (see p.72). These landslips extend from the plateau edge to the sea between Red Bay

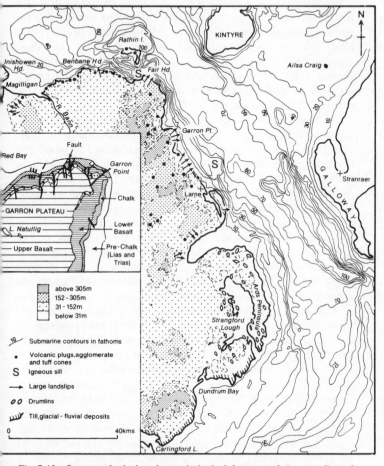

Fig. 7.13 Some geological and morphological features of the coastline of northeastern Ireland. An inset shows the detailed geological relationships at Garron Point where multiple faults (heavy black lines) have been picked out by deep ravines, and where large rotational landslips have occurred, probably during the latter part of the Devensian Cold Period. The landslips are almost certainly younger than 17 000 years B.P.

[D243257] and Belfast Lough, but except at Garron Point no sea-cliffs of any consequence occur. The outline of the slumped blocks has been softened by subsequent weathering and erosion; the seaward slopes carry a mantle of glacial drift; and between Carnlough and Larne mudflows carry large quantities of the debris on to the Antrim Coast Road, which, with its defensive walls, provides over long distances a semi-artificial shoreline (Prior *et al.* 1968; Hutchinson *et al.* 1974; fig. 7.12). Beaches of sand are restricted to the bay-heads, but massive basaltic boulder-beaches of limited width can be seen below the seawall protecting the Coast Road, as, for example, near Minnis North [D345127].

The North Channel coast is generally rocky, with stepped profiles predominating on the basalts and chalk. Abrasion platforms more than 19 m wide are uncommon, but can be observed on basalt and chalk at the northern end of Island Magee, on Keuper Marl at Carnlough, and on Triassic strata in Red Bay. 'Dead' cliffs abound, showing elevated notches and caves at their base, and are cut in basalt, chalk and Triassic conglomerate. These features range up to 8 m above mean sea-level and they represent mainly postglacial shorelines and notches, the elevated caves being particularly well preserved at Cushendun (in Old Red sandstone conglomerate) and at Red Bay (in Triassic strata). Excellent notches are seen in chalk along the Coast Road at Garron Point and just north of Larne; good elevated caves and notches in basalt can be seen at the southern end of The Gobbins cliffs on Island Magee. The rock-cut notches and caves are supplemented by good notches cut in till and in the regolith on the forward slopes of some of the slumped blocks (fig. 7.14). In addition, massive elevated shingle-bars are present at up to 8 m above mean sea-level at Cushendun (Movius 1940a), at 8-10 m above mean sea-level near Carnlough (Prior 1966), at Glenarm, and at Larne (Movius 1942). The age of the shingle-bars has been reviewed recently by Mitchell and Stephens (1974; see also Stephens and Synge 1966a, 1966b; Stephens 1968). There seems to be little doubt that isostatic recovery has been responsible for the elevation of many of the shingle-bars and their included Mesolithic artefacts, but there is the possibility of a fresh eustatic transgression at about 5000 years B.P., which could account for some of the bars. Further sites require detailed investigation, including the thick organic bed which can be observed in a river bank on Cushendall golf course below several metres of storm shingle. At Carnlough (between Glenarm and

Garron Point — fig. 7.2, p.184) an inter-tidal sequence of late- and postglacial sediments is known, similar to that at Roddans Port (Prior and Holland 1975).

Belfast Lough forms a major break in the continuity of the North Channel coastline. It is relatively shallow, being only about 20 m deep at its mouth, but the glacial drifts and other late- and postglacial marine sediments which choke both the lower Lagan valley, and the lough, conceal a buried channel in rock extending to at least 70 m below datum (see p.98). The postglacial marine 'estuarine clays' (Praeger 1893) are extensive at the head of the

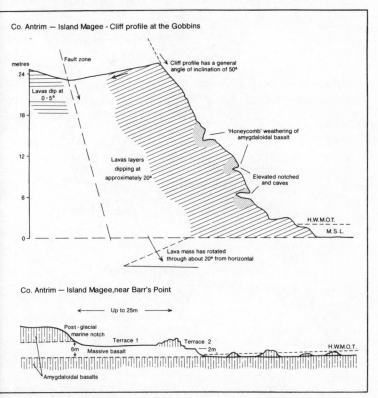

Fig. 7.14 Two profiles illustrating cliff forms at Island Magee. Near Barr's Point elevated notches and abrasion platforms are preserved in a sequence of 'massive' basalts lying between 'weaker' amygdaloidal basalts.

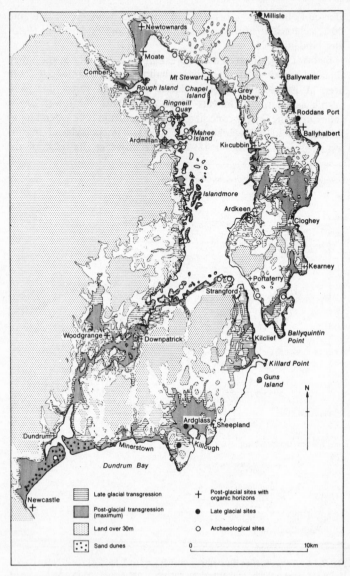

Fig. 7.15 Eastern Co. Down: Some aspects of the topography and the extent of the Late- and Post-Glacial marine transgressions.

lough, and in those parts of Belfast lying below 8 m above mean sea-level, where they are intercalated with organic horizons; at Castle Arcade (+3 m O.D.) timber near the base of the estuarine clay gave a date of 8715 ± 200 years B.P., and peat below the estuarine clay 9130 ± 120 years B.P. (Wilson, personal communication). Up to 12 m of 'estuarine clays' and organic horizons have been recorded, probably indicating progressive submergence as the sea flooded the lough after about 8400 years B.P.

The coastline of Co. Down, between Belfast Lough and Carlingford Lough, is formed by the partially drowned eastern edge of the Longford — Down axis, which gains in altitude only slowly away from the Irish Sea. Faulting may again account for sharp changes in general direction of the coast — for the north-to-south orientation of Strangford Lough, and for the northwest to southeast alignment of Carlingford Lough. The latter lough has been severely ice-scoured and over-deepened between Spelga and Carlingford Mountain, although the mouth of the lough is still obstructed by reefs developed in the Carboniferous limestone.

Much of the Palaeozoic bedrock is, nevertheless, hidden by glacial drifts and it is these which provide much of the variety to the coastal scenery. Drumlins composed of one or two till units make up the greater part of the coastline both within Strangford Lough, and along the outer coast of the Ards peninsula, where many drumlins are linked together by beach bars, as for example at Ballyhalbert [J645638] and Cloghey [J637562]. Countless drumlins have been removed by wave-attack and the Irish Sea coast has been 'smoothed' in the process. But inside Strangford Lough, with its restricted reef-strewn entrance channel between Portaferry and Strangford where seven-knot tidal-currents occur, many drumlins remain in groups or as single islands. Many others have been reduced to piles of boulders, cobbles, and gravel known locally as 'pladdies' and revealed only at low-tide.

To the south of the entrance to Strangford Lough, drumlins are absent, and morainic and outwash sands and gravels are truncated by cliffs up to 20 m high between Killard Point and Ardglass (fig. 7.15), with an ice-moulded abrasion platform outcropping in places at the base of the cliffs. The rock-platform is particularly well exposed at the entrance to Killough Harbour. Dundrum Bay forms a large indentation in the coastline where north-going and south-going tidal-currents meet. The bay shallows gradually, and is backed by extensive sand-dune systems which exceed 30 m in height within the Murlough Nature Reserve. The greater part of the older dunes rest upon, and largely obscure, a number of shingle-bars, some of which rise to over 8 m above mean sea-level (fig. 7.16). The progressive growth of these bars and hooked spits has enclosed Dundrum Inner Bay and the former lagoon areas between the nature reserve and Maghera.

The remainder of the Co. Down coastline flanks firstly the Mourne Mountains near Newcastle, where bevelled rock-cliffs reach heights of only about 10 to 12 m, and secondly the Mourne coastal plain, which is composed of a variety of glacial drifts. Cliffs up to 20 m high have been cut in hummocky morainic topography between Bloody Bridge [J390270] and Dunmore Head [J389244] and also to the south of Kilkeel. Between these points older till units and outwash deposits produce only a subdued morphology, but cliff heights may still exceed 20 m. Wave-attack is constantly under-cutting these relatively unresistant deposits, maintaining nearly vertical cliff profiles, although the precise cliff morphology varies greatly and is related as much to sub-aerial processes as to direct wave-action. Measurements to determine the rate of cliff-retreat have been conducted for several years at sites between Ballymartin [J340162] and Carlingford Lough. Preliminary results suggest that recession along this coastal segment is about 0.3 m per year, although this is greatly exceeded in places. Extensive areas of foreshore are strewn with thousands of large granite boulders derived from till sheets as the cliffs have been worn back. Exactly the same kind of boulder-beach can be seen near Rathcor [J183050] on the Carlingford peninsula, and it is apparent that general cliff-retreat has been sufficiently rapid to bring about the complete removal of those elevated postglacial notches and shorelines which elsewhere can be seen cut in these deposits.

Late-glacial and postglacial elevated shorelines and associated deposits are well exposed on the Co. Down coast, and have been the subject of considerable investigation in the last hundred years (Praeger 1896; Movius 1940b; Stephens and Collins 1960; Stephens and Synge 1966a). Many of the coastal drumlins are notched at 15-18 m (late-glacial) and at 4-8 m (postglacial) above mean sea-level, and the latter beach deposits have been identified by their shell content and lack of cryoturbation. Associated with the late-glacial shorelines are deposits of red clay and red sandy-silt, which can be

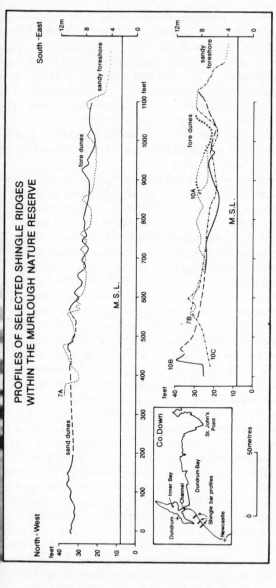

Fig. 7.16 Murlough Nature Reserve on Dundrum Bay, where shingle bars outcrop in hollows among the extensive sand-dune systems. A selection of profiles illustrates the elevated nature and varying altitude of some of the shingle bars. It is in the dune sands resting upon the shingle-bars that relic soils occur containing both archaeological material and charcoal. The latter has been dated by the radiocarbon method. Redrawn from a diagram by Mitchell and Stephens.

examined in section at a number of sites in the Ards peninsula, including Millisle [J600757], Roddans Port [J642645], Ballyquintin Point [J630453] and between Portaferry and Kircubbin. Well-marked washing-limits occur where the drumlin-forming tills have been eroded by wave-action to reveal the slates and grits of the Palaeozoic basement rocks (Stephens 1963). In 1965 details of Roddans Port, near Ballyhalbert, were first published, and it remains perhaps the most important site on the Co. Down coast, for here are exposed a sequence of deposits extending from Pollen Zone I to Pollen Zone VIc (about 13 000-6000 years B.P.), as shown in fig. 7.17 (Morrison and Stephens 1965).

Prior to Pollen Zone I, about 14 000 years ago, world sea-level was at about 70 m below datum. Consequently, the flooding of the inter-drumlin hollow at Roddans Port before or during Zone I indicates that the area was then deeply depressed below its modern level. Red marine clay accumulated in sheltered conditions among the drowned drumlins, notches were cut, beaches constructed, and bedrock washed clean of till, all at levels some 80 m below that of the present. Freshwater organic deposits of Zone II age occur at 57 m below datum in Bantry Bay (Stillman 1968), where little isostatic depression and recovery took place, and world sea-level was certainly lower than -57 m in Zone II. But at Roddans Port the first late-glacial marine transgression had taken place before Zone II, thus indicating the very considerable isostatic depression of the northeastern coast as compared with the southwestern peninsulas of Munster. Based upon the available evidence, the total amount of isostatic recovery in Co. Down since Zone I (c. 14 000 years B.P.) must be not less than 80 m (Stephens *et al.* 1975).

A tentative relationship has been established between the rate of downwasting and retreat of the ice-fronts in northern Ireland and southwestern Scotland, and the successive position of the elevated beaches and shorelines (Jardine 1964, 1971; Gemmell 1973). The phased withdrawal of the ice, and the associated marine transgressions and regressions (fig. 7.18) have been the subject of considerable investigation in the North Channel and the Firth of Clyde (Nichols 1967; Moar 1969; Bishop and Dickson 1970; Peacock 1971). It has been established that many of the older late-glacial shorelines are not synchronous features, even though they are found at the same heights. They are, in fact, metachronous, or time-transgressive, regardless of the height at which they are found (Stephens and Synge

1966a, 1966b; Stephens 1968). There is little doubt that further investigations on both sides of the North Channel will further elucidate the history of these shorelines in relation to the stages of ice-retreat.

The relative fall in sea-level which followed the marine transgression

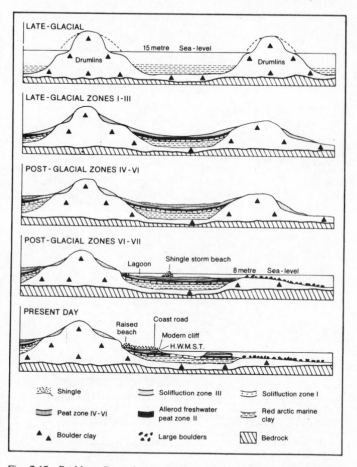

Fig. 7.17 Roddans Port, in the Ards peninsula, where a sequence of deposits in a former inter-drumlin hollow has provided useful evidence of Late- and Post-Glacial changes of land and sea level. Redrawn from a diagram by Morrison and Stephens.

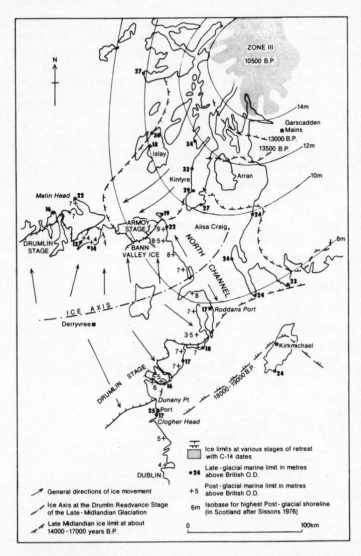

Fig. 7.18 A tentative chronology for ice retreat in the northern Irish Sea, North Channel, and adjacent sea and land areas. The Late-Glacial Marine Limit is shown for a selection of coastal sites to illustrate the relationship of the maximum of the transgression to the dated ice-limits in different places. With acknowledgement to G.F. Mitchell, F.M. Synge and A.M. McCabe.

in Zone I extended throughout Pollen Zones II to V in north-eastern Ireland, with respect to present mean sea-level. But subsequently there was a dramatic eustatic rise of sea-level which resulted in the transgression of various peat- and forest-beds in Lough Foyle, in Belfast Lough and in Strangford Lough, as well as at a host of other coastal sites (see fig. 7.15, p.212). On the coast of northeastern Ireland the oldest of the known drowned peat-beds have been dated at 8120 ± 135 years B.P. at Ballyhalbert, Co. Down, (Morrison and Stephens 1960), and organic horizons of similar age are known at Roddans Port, below the estuarine clays of Belfast Lough (Singh and Smith 1966), and in southwestern Scotland near Dumfries and Newton Stewart (Jardine 1964, 1971). The precise date of the marine transgression depends, of course, upon both the height of the particular organic horizon, and the ease with which the postglacial sea could penetrate the drift-covered landscape of kame moraine in Lough Foyle, or the drumlins occupying much of what is now Strangford Lough.

It has been demonstrated that there is a possibility that some of the marine deposits resting upon the organic horizons may be separated from them by a considerable time interval (Mitchell and Stephens 1974). In fig. 7.19 the situation at several sites is illustrated. At Woodgrange, a very sheltered site near Downpatrick, estuarine clay standing 1 m above mean sea-level has been dated at 6550 ± 300 years B.P., and that standing 1.5 m above mean sea-level at 3380 ± 180 years B.P. (Singh and Smith 1973). But at 3.5 m above mean sea-level marine sand is sealed by organic material dated at 3725 ± 150 years B.P., and regression from this maximum level may have begun about 3000 years B.P. In other words, the maximum of the marine transgression was not about 6500 years B.P. as might be suggested by the age of the lower estuarine clay.

At Ringneill Quay (see fig. 7.15, p.212) a similar hiatus in both the stratigraphy and the sequence of radiocarbon dates has been observed. Estuarine silt reached 2.5 m above mean sea-level and is dated at 7500 to 7345 ± 150 years B.P. A Neolithic occupation level, dated at 5380 ± 120 years B.P., rests upon the estuarine silt, and was subsequently transgressed by the sea. The upper marine series of shelly-sands and storm-shingle-beach is overlain in one section by a hearth (charcoal dated at 3680 ± 120 years B.P.), and a shell-midden (shells dated at 2660 ± 110 years B.P.) sealing the hearth. There is therefore certain evidence only that the maximum of the marine transgression was later than 5200, and younger than 3500, years B.P., suggesting that a regression occurred at about the same time as that at Woodgrange.

At present the evidence suggests that flooding of low-lying areas (less than 2.5 m above present mean sea-level) among the Co. Down drumlins may have taken place up to about 6500 years B.P., and allowed considerable penetration of the sea on the exposed Irish Sea

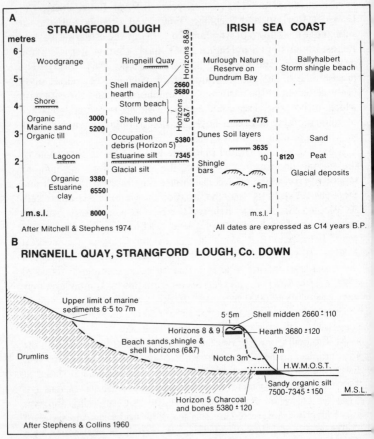

Fig. 7.19 Post-Glacial events in Co. Down. The relative levels of a series of sites are shown where 14C dated horizons have been recorded. Redrawn from a diagram by Mitchell and Stephens.

coast where wave-heights were greatest. But between 5000 and 3700 years B.P. an eustatic transgression allowed higher storm-beaches to be constructed, the erosion in the drumlins of those notches which still remain sharp morphological features, and permitted a much greater area to be flooded by a combination of salt and brackish water (see fig. 7.15, p.212). Associated with this final transgression are the Mesolithic and Neolithic remains.

The influence of exposure on the type and morphology of elevated postglacial shorelines is well demonstrated in Murlough Nature Reserve on Dundrum Bay. This bay is fully exposed to severe gales from the east and southeast, and below the dune systems there has been detected a series of magnificent storm-shingle-ridges. Some of these ridges are massive, and reach heights of 9 m above mean sea-level (see fig. 7.16, p.215). The general seaward trend of declining ridge height may reflect marine regression although this appears to have been abruptly reversed by a late phase of ridge construction, now below some of the existing foredunes. Unfortunately, it has not yet proved possible to date the most seaward of the shingle-ridges, although it is interesting that few, if any, archaeological finds have been made from the foredunes. But elsewhere within the Nature Reserve, Neolithic sites and buried soil-horizons stratified in the dune-sands above the shingle bars, have yielded dates of 4775 ± 140 years B.P. and 3635 ± 80 years B.P. Thus, the whole shingle-ridge complex may be no older than about 5000 years B.P., and could be attributed to a late eustatic transgression, or equally, only the foremost shingle-ridge lying below the foredunes may be that young.

While the shoreline diagrams which have been published for the northern and eastern coasts of Ireland seem to accommodate the known facts concerning the late- and postglacial shorelines reasonably well, there remain important problems to be solved (Stephens and McCabe 1977). The age of the late-glacial beaches and shore-lines, and their relationship to the phased withdrawal of the last ice-masses in northeastern Ireland both call for study. The northern Irish Sea and North Channel require further investigation, and there is as yet no firm chronology relating Ireland to the events already established for the Isle of Man (Dickson *et al*. 1970) and for south-western Scotland (Jardine 1971; Gemmell 1973). The precise age of the series of postglacial elevated notches and beaches (Stephens and Synge 1966a; Prior 1966; Mitchell and Stephens 1974) needs further clarification, identification and dating.

References

Akaad, M.K. (1956), The Ardara granitic diaper of county Donegal, Ireland. *Q.J. Geol. Soc. Lond.* 112, 263-90.

Anderson, T.B. (1965), The evidence for the Southern Uplands Fault in north-east Ireland. *Geol. Mag.* 102, 383-92.

Bamford, S.A.D. and Blundell, D.J. (1971), South-west Britain continental margin experiment, pp. 143-56 in Frances Margaret Delany (editor) *The geology of the East Atlantic continental margin,* Institute of Geological Sciences Report no. 70/14, H.M.S.O. London.

Batterbee, R.W. (1978), Observations on the recent history of Lough Neagh and its drainage basin. *Phil. Trans. R. Soc.* Series B 281, 304-420.

Bishop, W.W. and Dickson, J.H. (1970), Radiocarbon dates related to the Scottish late-glacial sea in the Firth of Clyde. *Nature* 227, 480-2.

Blundell, D.J., *et al.* (1968), Sedimentary basin in the south Irish Sea. *Nature* 219, 55-6.

Blundell, D.J., *et al.* (1971), Geophysical surveys over the south Irish Sea and Nymphe Bank. *Q. J. Geol. Soc. Lond.* 127, 339-75.

Bott, M.H.P. (1968), The geological structure of the Irish Sea basin. In Desmont Thomas Donovan (ed.), *Geology of shelf seas,* 93-115. Edinburgh.

Bowen, D.Q. (1973), The Pleistocene succession of the Irish Sea. *Proc. Geol. Ass.* 84, 249-72.

Brennand, T.P. (1965), The Upper Carboniferous (Namurian) stratigraphy north-east of Castleisland, co. Kerry, Ireland. *Proc. R. Ir. Acad.* 64B (4), 41-63.

Brindley, J.C. (1954), The geology of the northern end of the Leinster granite: part I — internal structural features. *Proc. R. Ir. Acad.* 56B (5), 159-190.

Brindley, J.C. (1973), The structural setting of the Leinster granite, Ireland: a review. *Scient. Proc. R. Dubl. Soc.* Series A 5(2), 27-36.

Brodrick, H. (1909), The Marble Arch caves, county Fermanagh: main stream series. *Proc. R. Ir. Acad.* 27B (9), 183-92.

Brown, P.E. and Miller, J.A. (1963), An absolute age determination on the Mourne Mountain Granite. *Geol. Mag.* 100, 93.

Bryant, R.H. (1966), The 'pre-glacial' raised beach in south-west Ireland. *Ir. Geogr.* 5(3), 188-203.

Bryant, R.H. (1968), *A study of the glaciation of South Iveragh, co. Kerry.* Ph.D. dissertation, University of Reading.

Bryant, R.H. (1974), A late-Midlandian section at Finglas River, near Waterville, Kerry. *Proc. R. Ir. Acad.* 74B (12), 161-78.

Burke, K. (1957), An outline of the structure of the Galway Granite. *Geol. Mag.* 94, 452-64.

Capewell, J.G. (1956), The stratigraphy, structure and sedimentation of the Old Red Sandstone of the Comeragh Mountains and adjacent areas, county Waterford, Ireland. *Q. J. Geol. Soc. Lond.* 112, 393-412.

Carter, R.W.G. (1975), Recent changes in the coastal geomorphology of the Magilligan Foreland, co. Londonderry. *Proc. R. Ir. Acad.* 75B (24), 469-97.

Chapman, R.J. (1970), The late-Weichselian glaciations of the Erne basin. *Ir. Geogr.* 6(2), 153-61.

Charlesworth, J.K. (1924), The glacial geology of the north-west of Ireland. *Proc. R. Ir. Acad.* 36B (12), 174-314.

Charlesworth, J.K. (1928a), The glacial geology of north Mayo and west Sligo. *Proc. R. Ir. Acad.* 38B (6), 100-15.

Charlesworth, J.K. (1928b), The glacial retreat from central and southern Ireland. *Q. J. Geol. Soc. Lond.* 84, 293-344.

Charlesworth, J.K. (1929), The glacial retreat in Iar Connacht. *Proc. R. Ir. Acad.* 39B (2), 95-106.

Charlesworth, J.K. (1931), The eskers of Ireland; their distribution, origin and human significance. *Geography* 16, 21-7.

Charlesworth, J.K. (1937), A map of the glacier-lakes and the local glaciers of the Wicklow Hills. *Proc. R. Ir. Acad.* 44B (3), 29-36.

Charlesworth, J.K. (1939), Some observations on the glaciation of north-east Ireland. *Proc. R. Ir. Acad.* 45B (11), 255-95.

Charlesworth, J.K. (1953), *The geology of Ireland: an introduction.* Edinburgh and London.

Charlesworth, J.K. (1963a), *Historical geology of Ireland.* Edinburgh and London.

Charlesworth, J.K. (1963b), Some observations on the Irish Pleistocene'. *Proc. R. Ir. Acad.* 62B (18), 295-322.

Charlesworth, J.K. (1963c), The bathymetry and origin of the larger lakes of Ireland. *Proc. R. Ir. Acad.* 63B (3), 61-9.

Charlesworth, J.K. (1973), Stages in the dissolution of the last ice-sheet in Ireland and the Irish Sea region. *Proc. R. Ir. Acad.* 73B (5), 79-86.

Close, M.H. (1867), Notes on the general glaciation of Ireland. *J. R. Geol. Soc. Ir.* 1, 207-42.

Cole, G.A.J. (1908), Probable Cretaceous and Cainozoic outliers off the coast of co. Kerry. *Geol. Mag.* n.s. Dec. 5, vol. 5, 463-4.

Cole, G.A.J. (1912), The problem of the Liffey valley. *Proc. R. Ir. Acad.* 30B (2), 8-19.

Cole, G.A.J. and Crook, T. (1910), *On rock-specimens dredged from the floor of the Atlantic off the coast of Ireland, and their bearing on submarine geology.* Memoirs of the Geological Survey of Ireland, Dublin.

Cole, G.A.J., *et al.* (1912), *The interbasaltic rocks (iron ores and bauxites) of north-east Ireland.* Memoirs of the Geological Survey of Ireland, Dublin.

Coleman, J.C. (1950), The Aille River and cave, co. Mayo, *Ir. Geogr.* 2(2), 58-60.

Coleman, J.C. (1952), Caves near Carrickmacross, county Monaghan. *Ir. Geogr.* 2(4), 180-3.

Coleman, J.C. (1955), Caves in the Cong area of Galway and Mayo. *Ir. Geogr.* 3(2), 94-106.

Coleman, J.C. (1965), *The caves of Ireland.* Tralee.

Coleman, J.C. and Dunnington, N.J. (1944), The Pollnagollum cave, co. Clare. *Proc. R. Ir. Acad.* 50B (5), 105-32.

Colhoun, E.A. (1967), Slope profiles of the Mourne Mountains, co. Down. *Ir. Geogr.* 5(4), 311-19.

Colhoun, E.A. (1970), On the nature of the glaciations and final deglaciation of the Sperrin Mountains and adjacent areas in the north of Ireland. *Ir. Geogr.* 6(2), 162-85.

Colhoun, E.A. (1971a), The glacial stratigraphy of the Sperrin Mountains and its relation to the glacial stratigraphy of north-west Ireland. *Proc. R. Ir. Acad.* 71B (2), 37-52.

Colhoun, E.A. (1971b), Late Weichselian periglacial phenomena of the Sperrin Mountains, Northern Ireland. *Proc. R. Ir. Acad.* 71B (3), 53-71.

Colhoun, E.A. (1972), The deglaciation of the Sperrin Mountains and adjacent areas in counties Tyrone, Londonderry and Donegal, Northern Ireland. *Proc. R. Ir. Acad.* 72B (8), 91-137.

Colhoun, E.A. (1973), Two Pleistocene sections in south-western Donegal and their relation to the last glaciation of the Glengesh plateau. *Ir. Geogr.* 6 (5), 594-609.

Colhoun, E.A., Common, R. and Cruickshank, M.M. (1965), Recent bog flows and debris slides in the north of Ireland. *Scient. Proc. R. Dubl. Soc.* Series A 2(10), 163-74.

Colhoun, E.A., Dickson, J.H., McCabe, A.M. and Shotton, F.W. (1972), A Middle Midlandian freshwater series at Derryvree, Maguiresbridge, county Fermanagh, Northern Ireland. *Proc. R. Soc.* B 180, 273-92.

Colhoun, E.A. and McCabe, A.M. (1973), Pleistocene glacial, glaciomarine and associated deposits of Mell and Tullyallen townlands, near Drogheda, eastern Ireland. *Proc. R. Ir. Acad.* 73B (12), 165-206.

Colhoun, E.A. and Mitchell, G.F. (1971), Interglacial marine formation and lateglacial freshwater formation in Shortalstown townland, co. Wexford. *Proc. R. Ir. Acad.* 71B (15), 211-45.

Colhoun, E.A., Ryder, A.T. and Stephens, N. (1973), C-14 age of an oak-hazel forest bed at Drumskellan, co. Donegal, and its relation to Late Midlandian and Littletonian raised beaches. *Ir. Nat. J.* 17, 321-7.

Corbel, J. (1957), *Les karsts du nord-ouest de l'Europe et de quelques regions de comparaison*. Institut des études Rhodaniennes de l'Université de Lyon: mémoires et documents 12, publication hors série de la Revue de Géographie de Lyon.

Coudé, A. (1973), Le district de volcanique Limerick (République d'Irlande): étude géomorphologique. *Norois* 79, 415-37.

Creighton, J.R. (1974), *A study of the late Pleistocene geomorphology of north-central Ulster*. Ph.D. dissertation, Queen's University, Belfast.

Crofts, R.S. (1971), Sand movement in the Emlybeg dunes, co. Mayo. *Ir. Nat. J.* 17, 132-6.

Crofts, R.S. (1976), Carbon-14 datings for research on sand dune formation in county Mayo, Ireland. Unpublished typescript.

Davies, G.L. (1958), Irish erosion surfaces — a statistical analysis. *Advmt Sci., Lond.* 14 (56), 385-8.

Davies, G.L. (1960a), The age and origin of the Leinster mountain chain: a study of the evolution of south-eastern Ireland from the Upper Palaeozoic to the later Tertiary. *Proc. R. Ir. Acad.* 61B (5), 79-107.

Davies, G.L. (1960b), Platforms developed in the boulder clay of the coastal margins of counties Wicklow and Wexford. *Ir. Geogr.* 4 (2), 107-16.

Davies, G.L. (1966), Cyclic surfaces in the Roundwood basin, co. Wicklow. *Ir. Geogr.* 5 (3), 150-60.

Davies, G.L. (1970), The enigma of the Irish Tertiary. In Nicholas Stephens and Robin Edgar Glasscock (eds), *Irish geographical studies in honour of E. Estyn Evans*, 1-16, Belfast.

Davies, G.L. and Whittow, J.B. (1975), A reconsideration of the drainage pattern of counties Cork and Waterford. *Ir. Geogr.* 8, 24-41.

Dewey, J.F. and McKerrow, W.S. (1963), An outline of the geomorphology of Murrisk and north-west Galway. *Geol. Mag.* 100, 260-75.

Dickson, C.A., Dickson, J.H. and Mitchell, G.F. (1970), The Late-Weichselian flora of the Isle of Man. *Phil Trans. R. Soc.* 258, 31-79.

Dobson, M.R., *et al.* (1973), *The geology of the south Irish Sea.* Institute of Geological Sciences. Report No. 73/11, H.M.S.O. London.

Drew, D.P. (1973), Ballyglunin cave, co. Galway, and the hydrology of the surrounding area. *Ir. Geogr.* 6(5), 610-17.

Drew, D.P. (1975), Landforms of the Burren, county Clare. *Geographical Viewpoint* 4, 21-38.

Dunnington, N.J. and Coleman, J.C. (1950), Dunmore cave, co. Kilkenny. *Proc. R. Ir. Acad.* 53B (2), 15-24.

Dury, G.H. (1957), A glacially breached watershed in Donegal. *Ir. Geogr.* 3(4), 171-80.

Dury, G.H. (1958), Glacial morphology of the Blue Stack area, Donegal. *Ir. Geogr.* 3 (5), 242-53.

Dury, G.H. (1959), A contribution to the geomorphology of central Donegal. *Proc. Geol. Ass.* 70, 1-27.

Dury, G.H. (1964), Aspects of the geomorphology of Slieve League peninsula, Donegal. *Proc. Geol. Ass.* 75, 445-59.

Dwerryhouse, A.R. (1923), The glaciation of north-eastern Ireland. *Q. J. Geol. Soc. Lond.* 79, 352-422.

Eden, R.A., *et al.* (1971), The solid geology of the East Atlantic continental margin adjacent to the British Isles. In Frances Margaret Delany (ed.), *The geology of the East Atlantic Continental margin*, Institute of Geological Sciences, 111-28 Report No. 70/14, H.M.S.O. London.

Evans, A.L., Fitch, F.J. and Miller, J.A. (1973), Potassium-argon age determinations on some British Tertiary igneous rocks. *J. Geol. Soc. Lond.* 129, 419-43.

Evans, D. (1973), A shallow seismic survey in Lough Swilly and Trawbreaga Bay, co. Donegal. *Proc. R. Ir. Acad.* 73B (13), 207-16.

Eyles, V. (1952), *The composition and origin of the Antrim laterites and bauxites*. Memoirs of the Geological Survey, H.M.S.O. Belfast.

Farrington, A. (1927), The topographical features of the granite-schist junction in the Leinster chain. *Proc. R. Ir. Acad.* 37B (20), 181-92.

Farrington, A. (1929), The pre-glacial topography of the Liffey basin. *Proc. R. Ir. Acad.* 38B (9), 148-70.

Farrington, A. (1931), The Loo valley, co. Kerry. *Proc. R. Ir. Acad.* 40B (10), 109-20.

Farrington, A. (1934), The glaciation of the Wicklow Mountains. *Proc. R. Ir. Acad.* 42B (7), 173-209.

Farrington, A. (1936), The glaciation of the Bantry Bay district. *Scient. Proc. R. Dubl. Soc.* n.s. 21 (37), 345-61.

Farrington, A. (1938), The local glaciers of Mount Leinster and Blackstairs Mountain. *Proc. R. Ir. Acad.* 45B (3), 65-71.

Farrington, A. (1942), The granite drift near Brittas, on the border between county Dublin and county Wicklow. *Proc. R. Ir. Acad.* 47B (12), 279-91.

Farrington, A. (1944), The glacial drifts of the district around Enniskerry, co. Wicklow. *Proc. R. Ir. Acad.* 50B (6), 133-57.

Farrington, A. (1947), Unglaciated areas in southern Ireland. *Ir. Geogr.* 1(4), 89-97.

Farrington, A. (1949), The glacial drifts of the Leinster Mountains. *J. Glaciol.* 1, 220-5.

Farrington, A. (1951), Notes on the geomorphology of the Kinsale district. *Ir. Geogr.* 2 (3), 124-8.

Farrington, A. (1953a), The South Ireland Peneplane. *Ir. Geogr.* 2 (5), 211-17.

Farrington, A. (1953b), Local Pleistocene glaciation and the level of the snow line of Croaghaun Mountain in Achill Island, co. Mayo, Ireland. *J. Glaciol.* 2, 262-7.

Farrington, A. (1954), A note on the correlation of the Kerry-Cork glaciations with those of the rest of Ireland. *Ir. Geogr.* 3 (1), 47-53.

Farrington, A. (1957a), Geomorphological notes on the area around Skibbereen. *Ir. Geogr.* 3 (4), 192-3.

Farrington, A. (1957b), Glacial Lake Blessington. *Ir. Geogr.* 3 (4), 216-22.

Farrington, A. (1959), The Lee Basin. Part One: glaciation. *Proc. R. Ir. Acad.* 60B (3), 135-66.

Farrington, A. (1961), The Lee Basin: Part 2, the drainage pattern. *Proc. R. Ir. Acad.* 61B (14) 233-53.

Farrington, A. (1964), Raised beaches in Galway Bay. *Ir. Nat. J.* 14, 216-17.

Farrington, A. (1965a), The last glaciation in the Burren, co. Care. *Proc. R. Ir. Acad.* 64B (3), 33-9.

Farrington, A. (1965b), A note on the correlation of some of the glacial drifts of the south of Ireland. *Ir. Nat. J.* 15, 29-33.

Farrington, A. (1966a), The last glacial episode in the Wicklow Mountains. *Ir. Nat. J.* 15, 226-8.

Farrington, A. (1966b), The early-glacial raised beach in county Cork. *Scient. Proc. R. Dubl. Soc.* Series A 2 (13), 197-219.

Farrington, A. (1968), Suggestions towards a history of the Shannon. *Ir. Geogr.* 5 (5), 402-7.

Farrington, A. and Haughton, J.P. (1947), The North Bull Island, co. Dublin. *Ir. Nat. J.* 9 (2), 46-8.

Farrington, A. and Mitchell, G.F. (1973), Some glacial features between Pollaphuca and Baltinglass, co. Wicklow. *Ir. Geogr.* 6 (5), 543-60.

Farrington, A. and Stephens, N. (1964), The Pleistocene geomorphology of Ireland. In James Alfred Steers (ed.), *Field studies in the British Isles*, 445-61, London.

Farrington, A. and Synge, F.M. (1970), The eskers of the Tullamore district. In Nicholas Stephens and Robin Edgar Glasscock (eds),

Irish geographical studies in honour of E. Estyn Evans, 49-52, Belfast.

Finch, T.F. (1966), Slieve Elva, co. Clare — a nunatak? *Ir. Nat. J.* 15, 133-6.

Finch, T.F. (1971), Limits of Weichsel glacial deposits in the south Tipperary area. *Scient. Proc. R. Dubl. Soc.* Series B 3 (2), 35-41.

Finch, T.F. (1974), Glacial retreat of late Midlandian ice in the Slieve Bernagh. *Scient. Proc. R. Dubl. Soc.* Series A 5 (15), 253-63.

Finch, T.F. and Synge, F.M. (1966), The drifts and soils of west Clare and the adjoining parts of counties Kerry and Limerick. *Ir. Geogr.* 5 (3), 161-72.

Finch, T.F. and Walsh, M. (1973), Drumlins of county Clare. *Proc. R. Ir. Acad.* 73B (23), 405-13.

Flatrès, P. (1954), Une surface d'aplanissement prédévonienne en Irlande (péninsule de Corran, comté de Mayo). *Norois* 1 (3), 231-40.

Flatrès, P. (1957), La Péninsule de Corran, comté de Mayo, Irlande: étude morphologique. *Bull. Soc. Géol. minér. Bretagne* n.s. fasc. 1, 1-63.

Flint, R.F. (1930), The origin of the Irish 'eskers'. *Geogrl Rev.* 20, 615-30.

Fowler, A. (1957), Geological Survey of Northern Ireland. In *Summary of progress of the Geological Survey of Great Britain and the Museum of Practical Geology for the year 1956*, 47-9, H.M.S.O. London.

Fowler, A. and Robbie, J.A. (1961), *Geology of the country around Dungannon*. Memoirs of the Geological Survey of Northern Ireland, H.M.S.O. Belfast.

Gardiner, M.J. and Ryan, P. (1964), The Soils of Co. Wexford. *Soil Survey Bulletin no. 1, National Soil Survey of Ireland*, An Foras Taluntais, Dublin.

Garrard, R.A. and Dobson, M.R. (1974), The nature and maximum extent of glacial sediments off the west coast of Wales. *Marine Geol.* 16, 31-44.

Gemmell, A.M.D. (1973), The deglaciation of the Island of Arran, Scotland. *Trans. Inst. Br. Geogr.* 59, 25-39.

Geological Survey (1861), *Explanations to accompany sheets 188 and 189 of the maps of the Geological Survey of Ireland, illustrating parts of the counties of Cork and Waterford.* Dublin.

George, T.N. (1967), Landform and structure in Ulster. *Scott. J. Geol.* 3, 413-48.

George, T.N. (1974), Prologue to a geomorphology of Britain. In Eric Herbert Brown and Ronald S. Waters (eds), *Progress in geomorphology: papers in honour of David L. Linton*, 113-25, Institute of British Geographers. London.

Gill, W.D. (1962), The Variscan fold belt in Ireland. In Kenneth Coe (ed.), *Some aspects of the Variscan fold belt*, 49-64, Manchester.

Guilcher, A. (1957), Les surfaces d'erosion fossiles exhumées dans le nord de l'Irlande. *Annls Géogr.* 66, 289-309.

Guilcher, A. (1962a), Morphologie de la baie de Clew (comté de Mayo, Irlande). *Bull. Ass. Géogr. fr.* no. 303-4, 53-65.

Guilcher, A. (1962b), Observations sur la morphologie littorale de la presqu'ile de Mullet et de la Baie le Blacksod. Ministère de l'éducation nationale, Comité des Travaux Historiques et Scientifiques, *Bulletin de la Section de Géographie* 75, 151-75.

Guilcher, A. (1965), Drumlin and spit structures in the Kenmare River south-west Ireland. *Ir. Geogr.* 5(2), 7-19.

Guilcher, A. (1966), Les grandes falaises et mégafalaises des côtes sud-ouest et ouest de l'Irlande. *Annls Geogr.* 75, 26-38.

Guilcher, A. and King, C.A.M. (1961), Spits, tombolos and tidal marshes in Connemara and west Kerry, Ireland. *Proc. R. Ir. Acad.* 61B (17), 283-338.

Hallissy, T. (1923), *Barytes in Ireland.* Memoirs of the Geological Survey of Ireland, Dublin.

Hancock, J.M. (1961), The Cretaceous sytem in Northern Ireland. *Q. J. Geol. Soc. Lond.* 117, 11-36.

Hancock, J.M. (1963), The hardness of the Irish chalk. *Ir. Nat. J.* 14, 157-64.

Hannon, M.A. (1975), *The late Pleistocene geomorphology of the Mourne Mountains and adjacent lowlands.* M.A. dissertation, Queen's University, Belfast.

Harris, C.R. (1974), The evolution of North Bull Island, Dublin Bay. *Scient. Proc. R. Dubl. Soc.* Series A 5 (14), 237-52.

Hartley, J.J. (1938), The Dalradian rocks of the Sperrin Mountains and adjacent areas in Northern Ireland. *Proc. R. Ir. Acad.* 44B (8), 141-71.

Hartley, J.J. (1948), The post-Mesozoic succession south of Lough Neagh, co. Antrim. *Ir. Nat. J.* 9, 115-21.

Hill, A.R. (1968), *An analysis of the spatial distribution and origin of drumlins in north Down and south Antrim, Northern Ireland.* Ph.D. dissertation, Queen's University, Belfast.

Hill, A.R. (1970), The relationship of drumlins to the directions of ice movement in north co. Down. In Nicholas Stephens and Robin Edgar Glasscock (eds), *Irish geographical studies in honour of E. Estyn Evans*, 53-9, Belfast.

Hill, A.R. (1971), The formation and spatial distribution of drumlins in a portion of north-east Ireland, in relation to hypotheses of drumlin origin. *Geogr. Ann.* 53A, 14-31.

Hill, A.R. (1973), The distribution of drumlins in county Down. *Ann. Ass. Am. Geogr.* 63, 226-40.

Hill, A.R. and Prior, D.B. (1968), Directions of ice movement in north-east Ireland. *Proc. R. Ir. Acad.* 66B (6), 71-84.

Hill, C.A., Brodrick, H. and Rule, A. (1909), The Mitchelstown caves, co. Tipperary. *Proc. R. Ir. Acad.* 27B (11), 235-68.

Hinch, J. de W. (1913), The Shelly Drift of Glenulra and Belderrig, Co. Mayo. *Ir. Nat.* 22, 1-6.

Hoare, P.G. (1975), 44 Shanganagh, co. Dublin. In *The Quaternary of the Wicklow District*, 15-17, Field Guide, Quaternary Research Association.

Hoare, P.G. and Synge, F.M., The glacial stratigraphy in Shanganagh and adjoining townlands, south-east County Dublin. *Proc. R. Ir. Acad.*, forthcoming.

Horne, R.R. and MacIntyre, R. (1975), Apparent age and significance of Tertiary dykes in the Dingle peninsula, S.W. Ireland. *Scient. Proc. R. Dubl. Soc.* Series A 5(18), 293-9.

Hull, E. (1891), *The physical geology and geography of Ireland.* Second edn, London.

Hutchinson, J.N., Prior, D.B. and Stephens, N. (1974), Potentially dangerous surges in an Antrim mudslide. *Q. J. Engng Geol.* 7, 363-76.

Jardine, W.G. (1964), Post-glacial sea-levels in south-west Scotland. *Scott. Geogr. Mag.* 80, 5-11.

Jardine, W.G. (1971), Form and age of late Quaternary shore-lines and coastal deposits of south-west Scotland: critical data. *Quaternaria* 14, 103-14.

Jardine, W.G. (1975), Chronology of Holocene marine transgression and regression in south-western Scotland. *Boreas* 4, 173-96.

Jessen, K. (1949), Studies in late Quaternary deposits and flora-history of Ireland. *Proc. R. Ir. Acad.* 52B (6), 85-290.

Jessen, K., Andersen S. Th. and Farrington, A. (1959), The

interglacial deposit near Gort, co. Galway, Ireland. *Proc. R. Ir. Acad.* 60B (1), 1-77.

Jukes, J.B. (1862a), *The student's manual of geology.* Edinburgh.

Jukes, J.B. (1862b), On the mode of formation of some of the river-valleys in the south of Ireland. *Q. J. Geol. Soc. Lond.* 18, 378-403.

Jukes, J.B., *et al.* (1861), *Explanations to accompany sheets 176 and 177 of the maps of the Geological Survey of Ireland.* Dublin and London.

Kidson, C. and Tooley, M.J. (1977), *The Quaternary history of the Irish Sea,* Liverpool.

Kidson, C. and Wood, R. (1974), The Pleistocene stratigraphy of Barnstaple Bay. *Proc. Geol. Ass.* 85, 223-37.

Kilroe, J.R. (1888), Directions of ice-flow in the north of Ireland. *Q. J. Geol. Soc. Lond.* 44, 827-33.

Kilroe, J.R. (1907a), *A description of the soil-geology of Ireland, based upon Geological Survey maps and records, with notes on climate.* Department of Agriculture and Technical Instruction for Ireland, Dublin.

Kilroe, J.R. (1907b), The River Shannon: its present course and geological history. *Proc. R. Ir. Acad.* 26B (8), 74-96.

King, C.A.M. and Gage, M. (1961), The extent of glaciation in part of west Kerry. *Ir. Geogr.* 4 (3), 202-8.

Kinahan, G.H. (1878), *Manual of the geology of Ireland.* London.

King, C.A.M. (1965), Some observations on the beaches of the west coast of county Donegal. *Ir. Geogr.* 5 (2), 40-50.

Lamont, A. (1939), Some shore features in south-east Ireland. *Scott. Geogr. Mag.* 55, 317-31.

Lamplugh, G.W., *et al.* (1903), *The geology of the country around Dublin.* Memoirs of the Geological Survey of Ireland, Dublin.

Lamplugh, G.W., *et al.* (1904), *The geology of the country around Belfast.* Memoirs of the Geological Survey of Ireland, Dublin.

Lamplugh, G.W., *et al.* (1905), *The geology of the country around Cork and Cork Harbour.* Memoirs of the Geological Survey of Ireland, Dublin.

Lamplugh, G.W. *et al,* (1907), *The geology of the country around Limerick.* Memoirs of the Geological Survey of Ireland, Dublin.

Langridge, D. (1971), Limestone pavement patterns on the island of Inishmore co. Galway. *Ir. Geogr.* 6(3), 282-93.

Leake, B.E. (1963), The location of the Southern Uplands Fault in central Ireland. *Geol. Mag.* 100, 420-3.

Leedal, G.P. and Walker, G.P.L. (1954), Tear faults in the Barnesmore area, Donegal. *Geol. Mag.* 91, 116-20.

Lewis, C.A. (1967), The glaciation of the Behy valley, county Kerry. *Ir. Geogr.* 5 (4), 293-301.

Lewis, C.A. (1974), The glaciations of the Dingle peninsula, county Kerry. *Scient. Proc. R. Dubl. Soc.* Series A 5 (13), 207-35.

Lewis, C.A. (1976), The Knockmealdown Mountains: a glacial nunatak. *Ir. Geogr.* 9, 18-28.

Lewis, H.C. (1894), *Papers and notes on the glacial geology of Great Britain and Ireland.* London.

Linton, D.L. (1951), Problems of Scottish scenery. *Scott. Geogr. Mag.* 67, 65-85.

Linton, D.L. (1964a), Tertiary landscape evolution. In James Wreford Watson and John Brian Sissons (eds), *The British Isles: a systematic geography*, 110-30, London.

Linton, D.L. (1964b), Aspects of the Pleistocene history of north Mayo. *Ir. Geogr.* 5 (1), 42-7.

McAulay, I.R. and Watts, W.A. (1961), Dublin radiocarbon dates I. *Radiocarbon* 3, 26-38.

McCabe, A.M. (1969a), A buried head deposit near Lisnaskea, Co. Fermanagh, Northern Ireland. *Ir. Nat. J.* 16, 232-3.

McCabe, A.M. (1969b), The glacial deposits of the Maguiresbridge area, county Fermanagh, Northern Ireland. *Ir. Geogr.* 6 (1), 63-77.

McCabe, A.M. (1972), Directions of late-Pleistocene ice-flows in eastern counties Meath and Louth, Ireland. *Ir. Geogr.* 6 (4), 443-61.

McCabe, A.M. (1973), The glacial stratigraphy of eastern counties Meath and Louth. *Proc. R. Ir. Acad.* 73B (21), 355-82.

Mackinder, Sir H.J. (1907), *Britain and the British seas.* Second edn, Oxford.

McManus, J. (1967), The influence of Pleistocene glaciation on the geomorphology of eastern Murrisk, co. Mayo. *Scient. Proc. R. Dubl. Soc.* Series A 3 (2), 17-31.

McMillan, N.F. (1957), Quaternary deposits around Lough Foyle, Northern Ireland. *Proc. R. Ir. Acad.* 58B (9), 185-205.

McMillan, N.F. (1964), The mollusca of the Wexford gravels

(Pleistocene), south-east Ireland. *Proc. R. Ir. Acad.* 63B (15), 265-89.

Manning, P.I., *et al.* (1970), *Geology of Belfast and the Lagan valley*. Memoirs of the Geological Survey, H.M.S.O. Belfast.

Miller, A.A. (1938), 'Pre-glacial erosion surfaces round the Irish Sea Basin', *Proc. Yorks. Geol. Soc.,* 24, 31-59.

Miller, A.A. (1939), River development in southern Ireland. *Proc. R. Ir. Acad.* 45B (14), 321-54.

Miller, A.A. (1955), The origin of the South Ireland Peneplane. *Ir. Geogr.* 3(2), 79-86.

Mitchell, G.F. (1941), Studies in Irish Quaternary deposits. No. 3 — the reindeer in Ireland. *Proc. R. Ir. Acad.* 46B, 183-8.

Mitchell, G.F. (1948), Two inter-glacial deposits in south-east Ireland. *Proc. R. Ir. Acad.* 52B (1), 1-14.

Mitchell, G.F. (1951), The Pleistocene period in Ireland. *Dansk Geol. Foren.* 12, 111-14.

Mitchell, G.F. (1956a), An early kitchen-midden at Sutton, co. Dublin (Studies in Irish Quaternary deposits: no. 12). *J. Roy. Soc. Antiq. Ir.* 86, 1-26.

Mitchell, G.F. (1956b), Post-boreal pollen-diagrams from Irish raised-bogs (Studies in Irish Quaternary deposits: no. 11). *Proc. R. Ir. Acad.* 57B (14), 85-251.

Mitchell, G.F. (1960), The Pleistocene history of the Irish Sea. *Advmt. Sci. Lond.* 17 no. 68, 313-25.

Mitchell, G.F. (1962), Summer field meeting in Wales and Ireland. *Proc. Geol. Ass.* 73, 197-213.

Mitchell, G.F. (1963), Morainic ridges on the floor of the Irish Sea. *Ir. Geogr.* 4 (5), 335-44.

Mitchell, G.F. (1965), Littleton bog, Tipperary: an Irish vegetational record. *Geol. Soc. Am. Spec. Publ.* 84, 1-16.

Mitchell, G.F. (1970a), The Quaternary deposits between Fenit and Spa on the north shore of Tralee Bay, co. Kerry. *Proc. R. Ir. Acad.* 70B (6), 141-62.

Mitchell, G.F. (1970b), Some chronological implications of the Irish Mesolithic. *Ulster J. Archael.* 33, 3-14.

Mitchell, G.F. (1971), Fossil pingos in the south of Ireland. *Nature* 230, 43-4.

Mitchell, G.F. (1972), The Pleistocene history of the Irish Sea: second approximation. *Scient. Proc. R. Dubl. Soc.* Series A 4 (13), 181-99.

Mitchell, G.F. (1973), Fossil pingos in Camaross townland, co. Wexford. *Proc. R. Ir. Acad.* 73B (16), 269-82.

Mitchell, G.F. (1976), *The Irish Landscape*, London.

Mitchell, G.F. and Parkes, H.M. (1949), The giant deer in Ireland (Studies in Irish Quaternary deposits, no. 6). *Proc. R. Ir. Acad.* 52B (7), 291-314.

Mitchell, G.F., Penny, L.F., Shotton, F.W. and West, R.G. (1973), A correlation of Quaternary deposits in the British Isles. *Geol. Soc. Lond., Special Report No. 4*, 1-99.

Mitchell, G.F. and Stephens, N. (1974), Is there evidence for a Holocene sea-level higher than that of today on the coasts of Ireland? C.N.R.S., *Les méthodes quantitatives d'étude des variations du climat au cours du Pléistocène* 219, 115-25.

Moar, N.T. (1969), Late Weichselian and Flandrian pollen diagrams from south-west Scotland. *New Phytol.* 68, 433-67.

Monkhouse, R.A. (1964), The geological structure of north county Cork, Ireland. *Scient. Proc. R. Dubl. Soc.* Series A 2(2), 13-19.

Morris, P. (1974), A Tertiary dyke system in south-west Ireland. *Proc. R. Ir. Acad.* 74B (13), 179-84.

Morrison, M.E.S. and Stephens, N. (1960), Stratigraphy and pollen analysis of the raised beach deposits at Ballyhalbert, co. Down, Northern Ireland. *New Phytol.* 59, 153-62.

Morrison, M.E.S. and Stephens, N. (1965), A submerged late-Quaternary deposit at Roddans Port on the north-east coast of Ireland. *Phil. Trans. R. Soc.* Series B 249 no. 758, 221-55.

Movius, H.L. (1940a), An early post-glacial archaeological site at Cushendun, co. Antrim. *Proc. R. Ir. Acad.* 46C (1), 1-84.

Movius, H.L. (1940b), Report on a Stone Age excavation at Rough Island, Strangford Lough. *J. Roy. Soc. Antiq. Ir.* 70, 111-42.

Movius, H.L. (1942), *The Irish Stone Age; its chronology, development and relationships.* Cambridge.

Murphy, T. (1952), *Measurements of gravity in Ireland: gravity surveys of central Ireland.* Dublin Institute for Advanced Studies, Geophysical Memoirs no. 2, part 3, Dublin.

Murphy, T. (1962), Some unusual low Bouguer anomalies of small extent in central Ireland and their connection with geological structure. *Geophys. Prospect.* 10 (3), 258-70.

Murphy, T. (1966), Deep alteration of Carboniferous strata in the Midleton, co. Cork district as detected by gravity surveying. *Proc. R. Ir. Acad.* 64B (17), 323-34.

Naylor, D. (1965), Pleistocene and post-Pleistocene sediments in Dublin Bay. *Scient. Proc. R. Dubl. Soc.* Series A 2(11), 175-88.

Naylor, D. and Mounteney, S.N. (1975), *Geology of the north-west European continental shelf.* Volume 1, London.

Nevill, W.E. (1958), The Carboniferous knoll-reefs of east-central Ireland. *Proc. R. Ir. Acad.* 59B (14), 285-303.

Nevill. W.E. (1963), *Geology and Ireland.* Dublin.

Nichols, H. (1967), Vegetational change, shoreline displacement and the human factor in the late Quaternary history of south-west Scotland. *Trans. R. Soc. Edinb.* 67, 145-87.

O'Reilly, H. and Pantin, G. (1957), Some observations on the salt marsh formation in co. Dublin. *Proc. R. Ir. Acad.* 58B (5), 89-218.

Orme, A.R. (1964), Planation surfaces in the Drum Hills, county Waterford, and their wider implications. *Ir. Geogr.* 5 (1), 48-72.

Orme, A.R. (1966), Quaternary changes of sea-level in Ireland. *Trans. Inst. Brit. Geogr.* 39, 127-40.

Orme, A.R. (1967), Drumlins and the Weichsel glaciation of Connemara. *Ir. Geogr.* 5 (4), 262-74.

Orme, A.R. (1969), The Quaternary glaciation of Ireland. *Geogr. Viewpoint* 2 (1), 15-26.

Oswald, D.H. (1955), The Carboniferous rocks between the Ox Mountains and Donegal Bay. *Q. J. Geol. Soc. Lond.* 111, 167-86.

Peacock, J.D. (1971), Marine shell radiocarbon dates and the chronology of deglaciation in Western Scotland. *Nature, Physical Science* 230, 43-5.

Pilcher, J.R. (1969), Archaeology, palaeoecology and radiocarbon dating of the Beaghmore stone circle site. *Ulster J. Archael.* 32, 73-92.

Pitcher, W.S., *et al.* (1964), The Leannan Fault. *Q. J. Geol. Soc. Lond.* 120, 241-73.

Pitcher, W.S. and Berger, A.R. (1972), *The geology of Donegal: a study of granite emplacement and unroofing.* New York.

Praeger, R. Ll. (1893), Report on the estuarine clays of the north-east of Ireland. *Proc. R. Ir. Acad.* Series 3, 2, 212-89.

Praeger, R.Ll. (1896), Report upon the raised beaches of the north-east of Ireland with special reference to their fauna. *Proc. R. Ir. Acad.* 4, 30-54.

Prior, D.B. (1966), Late glacial and post-glacial shorelines in north-east Antrim. *Ir. Geogr.* 5(3), 173-87.

Prior, D.B. (1968), *The late-Pleistocene geomorphology of north-east Antrim*. Ph.D. dissertation, Queen's University, Belfast.

Prior, D.B. (1970), Ice limits in the Cushendun area of northeast co. Antrim. In Nicholas Stephens and Robin Edgar Glasscock (eds), *Irish geographical studies in honour of E. Estyn Evans*, 59-64, Belfast.

Prior, D.B. (1975), A mudslide on the Antrim coast, 24 November 1974. *Ir. Geogr.* 8, 55-62.

Prior, D.B. and Holland, S. (1975), A Late Quaternary sediment sequence at Carnlough, Co. Antrim. *Abst. Ir. Quat. Res. Meeting*, 4-5.

Prior, D.B., Stephens, N. and Archer, D.R. (1968), Composite mudflows on the Antrim coast of north-east Ireland. *Geogr. Ann.* 50A, 65-78.

Proudfoot, V.B. (1954), Erosion surfaces in the Mourne Mountains. *Ir. Geogr.* 3 (1), 26-35.

Reed, F.R.C. (1906), Notes on the corries of the Comeragh Mountains co. Waterford. *Geol. Mag.* n.s. Dec. 5, vol. 3, 154-61, 227-34.

Reffay, A. (1966), Problèmes morphologiques dans la péninsule de sud-ouest du Donegal. *Revue Géogr. alp.* 54 (2), 287-312.

Reffay, A. (1968), Mise au point sur les caractères physiques de l'Irlande. *Revue Géogr. alp.* 56 (1), 153-76.

Reffay, A. (1969), Une côte à falaises basaltiques: le promontoire de la Chaussée des Géants (comté d'Antrim, Irlande du Nord). *Revue Géogr. alp.* 57 (4), 783-801.

Reffay, A. (1972), *Les montagnes de l'Irlande septentrionale: contribution à la géographie physique de la montagne atlantique.* Grenoble.

Reynolds, D.L. (1961), Lapiés and solution pits in olivine-dolerite sills at Slieve Gullion, Northern Ireland. *J. Geol.,* 69, 110-17.

Richey, J.E. (1927), The structural relations of the Mourne granites (Northern Ireland). *Q. J. Geol. Soc. Lond.* 69, 653-88.

Rohleder, H.P.T. (1932), A tectonic analysis of the Mourne granite mass, county Down. *Proc. R. Ir. Acad.* 40B (12), 160-74.

Schou, A. (1945), Det marine forland. *Folia geogr. dan.* 4, 1-236.

Selwood, E.B. and Coe, K. (1963), Large-scale terminal curvature affecting the cliffs west of Castletown Berehaven, west Cork. *Proc. Geol. Ass.* 74 (4), 461-5.

Seymour, H.J. (1939), Bathymetric survey of three lakes in co. Wicklow. *Proc. R. Ir. Acad.* 45B (12), 297-9.

Singh, G. (1970), Late-glacial vegetational history of Lecale, co. Down. *Proc. R. Ir. Acad.* 69B (10), 189-216.

Singh, G. and Smith, A.G. (1966), The post-glacial marine transgression in Northern Ireland — conclusions from estuarine and 'raised beach' deposits: a contrast. *Palaeobotanist* 15, 230-4.

Singh, G. and Smith, A.G. (1973), Post-glacial vegetational history and relative land- and sea-level changes in Lecale, co. Down. *Proc. R. Ir. Acad.* 73B (1), 1-51.

Smith, A.G. (1961), Cannons Lough, Kilrea, Co. Derry: stratigraphy and pollen analysis. *Proc. R. Ir. Acad.* 61B, 369-83.

Smith, A.G. (1970), Late- and post-glacial vegetational and climatic history of Ireland: a review. In Nicholas Stephens and Robin Edgar Glasscock (eds), *Irish geographical studies in honour of E. Estyn Evans*, 65-88, Belfast.

Sollas, W.J. (1893-6), A map to show the distribution of eskers in Ireland. *Scient. Trans. R. Dubl. Soc*. Series 2, 5, 785-822.

Stelfox, *et al*. (1972), The late-glacial and post-glacial mollusca of the White Bog, Co. Down. *Proc. R. Ir. Acad.* 72B, 185-207.

Stephens, N. (1957), Some observations on the 'Interglacial' platform and the early post-glacial raised beach on the east coast of Ireland. *Proc. R. Ir. Acad.* 58B (6), 129-49.

Stephens, N. (1958), The evolution of the coastline of north-east Ireland. *Advmt. Sci.* 14 (56), 389-91.

Stephens, N. (1963), Late-glacial sea-levels in north-east Ireland. *Ir. Geogr*. 4 (5), 345-59.

Stephens, N. (1968), Late-glacial and post-glacial shorelines in Ireland and south-west Scotland. In *Means of correlation of Quaternary successions*, 437-56 Internat. Studies on the Quaternary, VII Congress Intern. Assoc. Quat. Research, Boulder, Colorado.

Stephens, N. (1970), The coastline of Ireland. In Nicholas Stephens and Robin Edgar Glasscock (eds), *Irish geographical studies in honour of E. Estyn Evans*, 125-45, Belfast.

Stephens, N. and Collins, A.E.P. (1960), The Quaternary deposits at Ringneill Quay and Ardmillan, Co. Down. *Proc. R. Ir. Acad.* 61C (3), 41-77.

Stephens, N., Creighton, J.R. and Hannon, M.A. (1975), The late-Pleistocene period in north-eastern Ireland: an assessment 1975. *Ir. Geogr.* 8, 1-23.

Stephens, N. and McCabe, A.M. (1977), Late-Pleistocene ice movements and patterns of late- and post-glacial shorelines on the coast of Ulster. In C. Kidson and M.J. Tooley (eds), *Quaternary History of the Irish Sea, Geol. J.*, special issue no. 7, 179-98.

Stephens, N. and Synge, F.M. (1958), A Quaternary succession at Sutton, co. Dublin. *Proc. R. Ir. Acad.* 59B (2), 19-25.

Stephens, N. and Synge, F.M. (1965), Late-Pleistocene shorelines and drift limits in north Donegal. *Proc. R. Ir. Acad.* 64B (9), 131-53.

Stephens, N. and Synge, F.M. (1966a), Late- and post-glacial shorelines, and ice limits in Argyll and north-east Ulster. *Trans. Inst. Br. Geogr.* 39, 101-25.

Stephens, N. and Synge, F.M. (1966b), Pleistocene shorelines. In George Harry Dury (ed.), *Essays in geomorphology*, 1-51, London.

Stillman, C.J. (1968), The post-glacial change in sea level in south-western Ireland: new evidence from fresh-water deposits on the floor of Bantry Bay. *Scient. Proc. R. Dubl. Soc.* Series A 3 (11), 125-7.

Sweeting, M.M. (1953), The enclosed depression of Carran, county Clare. *Ir. Geogr.* 2 (5), 218-24.

Sweeting, M.M. (1955), The landforms of north-west county Clare, Ireland. *Trans. Inst. Br. Geogr.* 21, 33-49.

Synge, F.M. (1950), The glacial deposits around Trim, co. Meath. *Proc. R. Ir. Acad.*, 53B (10), 99-110.

Synge, F.M. (1963a), A correlation between the drifts of south east Ireland and those of west Wales. *Ir. Geogr.* 4 (5), 360-6.

Synge, F.M. (1963b), The glaciation of the Nephin Beg range, co. Mayo. *Ir. Geogr.* 4(6), 397-403.

Synge, F.M. (1964), Some problems concerned with the glacial succession in south-east Ireland. *Ir. Geogr.* 5, 73-82.

Synge, F.M. (1968), The glaciation of west Mayo. *Ir. Geogr.* 5(5), 372-86.

Synge, F.M. (1969), The Würm ice limit in the west of Ireland. In *Quaternary geology and climate, Publ. 1701*, 89-92. Nat. Acad. Sciences, Washington, D.C., 89-82.

Synge, F.M. (1970), The Irish Quaternary: current views 1969. In Nicholas Stephens and Robin Edgar Glasscock (eds), *Irish geographical studies in honour of E. Estyn Evans*, 34-48, Belfast.

Synge, F.M. (1973), The glaciation of south Wicklow and the adjoining parts of the neighbouring counties. *Ir. Geogr.* 6 (5), 561-9.

Synge, F.M. (1977), The coasts of Leinster. In C. Kidson and M.J. Tooley (eds), *Quaternary History of the Irish Sea*, Geol. J., special issue no. 7.

Synge, F.M. and Stephens, N. (1960), The Quaternary period in Ireland — an assessment, 1960. *Ir. Geogr.* 4(2), 121-30.

Thirlaway, H.I.S. (1951), *Measurements of gravity in Ireland: gravimeter observations between Dublin, Sligo, Galway and Cork*. Dublin Institute for Advanced Studies Geophysical Memoirs, 2 (2), Dublin.

Thorp, M.B. (1962), *Erosion surfaces, drainage evolution, and denudation chronology of the Sperrin mountains, Northern Ireland*. M.A. dissertation, University of Liverpool.

Tratman, E.K. (ed) (1969), *The caves of north-west Clare, Ireland*. Newton Abbot.

Vernon, P. (1965), Implications of foreign or erratic stones on Slieve Donard, Mourne Mountains. *Irish Nat. J.* 15, 36-8.

Vernon, P. (1966), Drumlins and Pleistocene ice flow over the Ards peninsula/Strangford Lough area, county Down, Ireland. *J. Glaciol.* 6, (45), 401-9.

Walker, G.P.L. and Leedal, G.P. (1954), The Barnesmore granite complex, county Donegal. *Scient. Proc. R. Dubl. Soc.* n.s. 26 (13), 207-43.

Walsh, P.T. (1959-60), Specimens from an occurrence of Cretaceous chalk in the Killarney district, Eire. *Proc. Geol. Soc.* 1581, 112-13.

Walsh, P.T. (1965), Possible Tertiary outliers from the Gweestin valley, co. Kerry. *Ir. Nat. J.* 15 (4), 100-4.

Walsh, P.T. (1966), Cretaceous outliers in south-west Ireland and their implications for Cretaceous palaeogeography. *Q. J. Geol. Soc. Lond.* 122, 63-84.

Warren, W.P. (1970), *Cirque glaciation of county Dublin*. B.A. dissertation, National University of Ireland (University College, Dublin).

Watts, W.A. (1957), A Tertiary deposit in county Tipperary. *Scient. Proc. R. Dubl. Soc.* N.S. 27 (13), 309-11.

Watts, W.A. (1959), Interglacial deposits at Kilbeg and Newtown, co. Waterford. *Proc. R. Ir. Acad.*, 60B (2), 79-134.

Watts, W.A. (1963), Late-glacial pollen zones in western Ireland. *Ir. Geogr.* 4 (5), 367-76.

Watts, W.A. (1964), Interglacial deposits at Baggotstown, near

Bruff, co. Limerick. *Proc. R. Ir. Acad.,* 63B (9), 167-89.

Watts, W.A. (1967), Interglacial deposits in Kildromin townland, near Herbertstown, co. Limerick. *Proc. R. Ir. Acad.* 65B (15), 339-48.

Watts, W.A. (1970), Tertiary and interglacial floras in Ireland. In Nicholas Stephens and Robin Edgar Glasscock (eds), *Irish geographical studies in honour of E. Estyn Evans,* 17-33, Belfast.

Watts, W.A. (1971), The identity of *Menyanthes Microsperma* N. SP. Foss. from the Gort Interglacial, Ireland. *New Phytol.* 70, 435-6.

West, R.G. (1957), Interglacial deposits at Bobbitshole, Ipswich. *Phil. Trans. R. Soc.* 241, 1-31.

West, R.G. (1968), *Pleistocene geology and biology with especial reference to the British Isles.* London.

Whittow, J.B. (1958), The structure of the southern Irish Sea area. *Advmt. Sci.* 14 (56), 381-5.

Whittow, J.B. (1974), *Geology and scenery in Ireland.* Penguin books, Harmondsworth.

Wilkinson, S.B., *et al* (1908), *The geology of the country around Londonderry.* Memoirs of the Geological Survey of Ireland, Dublin.

Williams, M. (1964), Glacial breaches and sub-glacial channels in south-western Ireland. *Ir. Geogr.* 5 (1), 83-95.

Williams, P.W. (1963), An initial estimate of the speed of limestone solution in county Clare. *Ir. Geogr.* 4 (6), 432-41.

Williams, P.W. (1966), Limestone pavements with special reference to western Ireland. *Trans. Inst. Br. Geogr.* 40, 155-72.

Williams, P.W. (1968), An evaluation of the rate and distribution of limestone solution and deposition in the River Fergus basin, western Ireland. In *Contributions to the study of karst,* 1-40, Australian National University, Canberra, Research School of Pacific Studies, Department of Geography Publication G/5.

Williams, P.W. (1970), Limestone morphology in Ireland. In Nicholas Stephens and Robin Edgar Glasscock (eds), *Irish geographical studies in honour of E. Estyn Evans,* 105-24, Belfast.

Wilson, H.E. (1972), *Regional geology of Northern Ireland.* Geological Survey of Northern Ireland, H.M.S.O. Belfast.

Wilson, R.L. (1964), The Tertiary dykes of Magho Mountain, co. Fermanagh. *Ir. Nat. J.* 14 (11), 254-7.

Wright, P.C. (1964), The petrology, chemistry and structure of the Galway Granite of the Carna area, co. Galway. *Proc. R. Ir. Acad.* 63B (14), 239-64.

Wright, W.B. (1912), The drumlin topography of south Donegal. *Geol. Mag.* n.s. Dec. 5, vol. 9, 153-9.

Wright, W.B. (1914), *The Quaternary Ice Age.* London.

Wright, W.B. (1920), Minor periodicity in glacial retreat. *Proc. R. Ir. Acad.* 35B (6), 93-105.

Wright, W.B. (1924), Age and origin of the Lough Neagh clays. *Q. J. Geol. Soc. Lond.* 80, 468-88.

Wright, W.B. (1937), *The Quaternary Ice Age.* Second Edn, London.

Wright, W.B., *et al.* (1927), *The geology of Killarney and Kenmare.* Memoirs of the Geological Survey of Ireland, Dublin.

Wright, W.B. and Muff, H.B. (1904), The pre-glacial raised beach of the south coast of Ireland. *Scient. Proc. R. Dubl. Soc.* n.s. 10, part 2, no. 25, 250-324.

Index

Abbeyfeal Plateau, 41, 44
Achill Island, 51, 52, 136, 147, 148, 196
Aghavannagh, 26
Aille river, 21
Allen, Lough, 22
Altiplanation terraces, 48
Alts, 55
Annalong valley, 74, 162, 165
Annamoe, 109
Annamoe river, 109
Antecedent drainage, 102
Antrim, 5
Antrim Coast Readvance, 167
Antrim Coast Road, 5, 68, 71, 72, 210
Antrim Glens, 71, 96, 112, 167, 208
Antrim Plateau, 5-6, 9, 66-72, 96, 118, 167, 176
Aran Island, Co. Donegal, 203
Aran Islands, Co. Galway, 46, 193
Ardara, 62, 154, 155, 201
Ardglass, 168, 213
Ardmore Syncline, 107
Ards Peninsula, 1, 73, 170, 173, 213, 216
Arklow, 31, 128, 129, 187
Arklow Head, 26, 31, 125, 183
Arklow Rock, 31
Armorican structures, 4, 30, 32-9, 39-40, 103-7, 181
Armoy, 113, 167
Armoy Moraine, 71, 113, 118, 120, 122, 167, 173

Arra Mountains, 24, 101
Ashford, 29
Athdown Stage, 130, 133, 134, 145
Athy, 102
Aughrim, 29, 109, 129
Aughrim Advance, 129, 130
Aughrim-Kilvadin Corridor, 108, 109
Avoca, 129
Avonbeg Glacier, 129
Avonmore Glacier, 130

Baggotstown, 145
Balbriggan, 160
Ballina Syncline, 52, 55, 197
Ballinaclash, 130
Ballingarry Hills, 20
Ballon Hill, 94
Ballybrack, 129
Ballycastle, Co. Antrim, 68, 70, 71, 112, 113, 167, 173, 205
Ballycastle, Co. Mayo, 54, 147, 152, 155
Ballycastle-Mulrany Moraine, 147
Ballycotton, 92, 124, 140, 141, 189
Ballycotton Anticline, 36, 104
Ballycroneen till, 116, 129, 140, 141, 190
Ballydeenlea, Co. Kerry, 5, 81, 82, 85, 86-7
Ballydonnell Brook, 27
Ballyferriter, 192
Ballyhalbert, 204, 213, 216, 219
Ballyhoura Mountains, 40

Ballylanders Basin, 40, 143, 144
Ballylanders Moraine, 120, 136, 143, 144, 145, 165
Ballymacadam, 7, 80, 81, 82, 85, 88, 90, 98
Ballymartin Glaciation, 162
Ballysadare, 55, 110, 199
Ballysadare river, 110
Ballyvoyle till, 140
Baltinglass, Co. Wicklow, 27, 30
Bandon river, 32, 37, 104, 107
Bangor, Co. Down, 171
Bangor, Co. Mayo, 54, 110
Bann river, Co. Wexford, 109
Bann valley, Ulster, 122, 152, 154, 165, 173, 174, 176
Bannow till, 140
Bantry Bay, 38, 190, 192, 197, 216
Barnesmore, 154, 155
Barnesmore gap, 155
Barrow river, 30, 31, 41, 81, 94, 98, 100, 102, 103, 107, 108, 113, 136
Basaltic Plateau, 5-6, 9, 118, 66-72, 96, 167, 176
'Basket of eggs', 8, 73
Beara Peninsula, 7, 37, 38, 84, 143
Belderg, 52, 96, 147, 152, 197
Belfast, 5, 66, 98, 165, 174, 176, 213
Belfast Lough, 15, 72, 73, 98, 125, 126, 167, 208, 210, 211, 213, 219
Benbulbin Range, 56-9
Benwee Head, 51, 197, 198
Blacksod Bay, 52, 196, 197
Blackstairs Mountains, 29-31, 92, 94, 100
Blackwater, Co. Wexford, 31, 187
Blackwater river, Co. Waterford, 35, 91, 98, 103-7, 141
Blanket Nook Glaciation, 152, 154, 160
Blarney Syncline, 105
Blessington, Glacial Lake, 134
Blessington Moraine, 136
Blessington Reservoir, 27, 109, 134
Blue Stack Mountains, 62, 63, 64
Boggaragh Mountains, 34, 92, 105
Bowen, D.Q., 190
Boyne, Lord, 203

Boyne river, 17, 97
Brandon Hill, 4, 30, 94, 100
Brandon Mountain, 39
Bray, 136
Bray, loughs, 27, 133, 134, 136
Bray Head, 28, 29, 183
Bricklieve Mountains, 56-9
Brindley, J.C., 24
Brittas Bay, 187
Brittas Glaciation, 130, 134, 136
Broad Haven, 147, 196, 197
Bryant, R.H., 192
Bunclody Gap, 100, 103
Bunowen, 7, 48
Buried channels, 98-9, 188, 189, 192, 211
Burke, R., 95
Burnfoot till, 152·
Burren, 44-6, 193
Bush river, Co. Antrim, 71, 112-13, 205
Bushmills, 69, 71

Caha Mountains, 143
Caha Peninsula, 37
Cahore Point, 92, 126, 187
Caledonian structures, 1, 4, 24, 30, 49, 50, 54, 60, 65, 71, 72, 87-8, 95, 107, 108, 111, 201
Callows, 22
Canon's Lough, 174, 179
Cappoquin, 35, 103, 104, 107
Carey valley, 71, 166, 167
Carlingford Lough, 69, 78, 168, 183, 213, 214
Carlingford Mountains, 6, 11, 73, 77-8, 83, 112, 122, 160, 165
Carlingford Peninsula, 183, 185
Carnlough, 68, 210
Carnsore Point, 138, 187
Carran depression, 46
Carrauntoohil, 14, 38
Carrick-on-Shannon, 15
Carrick-on-Suir, 36
Carrick Syncline, 101
Cartron River Deposit, 147
Castlecomer, 41
Castlecomer Plateau, 30, 40-1, 42, 102, 122

Castlemartyr Syncline, 8, 98
Castlerock, 70, 204, 205
Cavan-Down Hill Country, 72-3
Central Lowland, 11, 17-24, 40, 41,
    42, 55, 56, 78, 85, 97-8, 100, 101,
    109, 113-14, 118, 122, 129, 148,
    165, 176-9, 195
Chapman, R.J. 149
Charlesworth, J.K., 115, 118, 122,
    152, 154, 157, 167
Cirques, 8, 27, 36, 38, 39, 52-4, 59,
    63, 64, 66, 74, 143, 145, 148, 149,
    162
Clare Island, 15, 196
Clare Plateau, 44-6, 93, 193
Clew Bay, 11, 14, 15, 49, 50, 51, 52,
    54, 97, 122, 147, 148, 195, 196,
    197
Cliffs of Moher, 44, 193
Climatic geomorphology, 13
Clogga, 125
Clogga Head, 187
Clogga till, 126, 129
Clonakilty, 91, 189
Close, M.H., 97, 115
Cloyne Syncline, 107
Coast erosion, 189, 190, 192, 193,
    196, 199, 201, 214, 216
Coastal Peneplane, 92, 93, 95, 96, 97
Colbinstown Stage, 134
Cole, G.A.J., 79
Coleraine, 69
Colhoun, E.A., 74, 134, 138, 156,
    157
Comeragh Mountains, 8, 35, 36, 92,
    105
Composite mudflows, 72, 210-11
Conn, Lough, 51, 52, 55
Conemara, 8, 116
Connemara Antiform, 49
Connemara Lowland, 15, 46-8
Consequent rivers, 101-3, 104, 111
Cork, 21, 36, 98, 104, 107
Cork Harbour, 32, 34, 92, 98, 104,
    105, 116, 141, 189
Cork-Kerry Ice-Cap, 8
Cork Syncline, 8, 104, 105
Corraun Interglacial, 147
Corraun Peninsula, 52, 95, 147, 196

Corrib, Lough, 17, 21, 22, 48, 50
Courtmacsherry Bay, 124, 189
Courtown, 187, 190
Creighton, J.R., 122
Cretaceous chalk cover, 79-83, 90
Croagh Patrick, 14, 50, 51, 146, 147,
    148, 196
Cuilcagh Mountains, 22, 56, 59, 84,
    149
Cuilcagh Plateau Country, 56-9, 199
Cummer Vale Ridge, 29, 31, 100,
    109
Cushendall, 15, 68, 210
Cushendun, 68, 210

Dargle river, 109, 133
Dartry Mountains, 56-9
Davies, G.L., 107
Deers Meadow, 74, 76, 162
Derg, Lough, 22, 24, 96, 101, 113
Derry Corridor, 108, 109, 126, 128
Derry river, 29, 109
Derry Water, 29, 109
Derryveagh Mountains, 60-2, 64
Derryvree, 154, 170
Devensian Glaciation, 8, 69
Devil's Glen, 94, 109
Dewey, J.F., 50, 87, 95, 110
Dingle Bay, 4, 141, 143, 190, 192
Dingle Peninsula, 7, 37, 38-9, 42, 84,
    145, 192
Donegal, 7, 63
Donegal-Ballina Lowland, 55-6
Donegal Bay, 55, 147, 152, 154, 155,
    168, 199, 200
Donegal Highlands, 59-64, 111
Donegal Syncline, 55
Doon Hill, 7, 48
Drainage superimposition, 24, 29,
    30, 54, 81, 85, 110, 111, 112, 113-
    14
Drew, D.P., 93
Drogheda, 4, 116, 160
Drogheda Glaciation, 160, 161
Dromana Forest Anticline, 35, 36
Drumlin Readvance, 120, 149, 155,
    156, 157, 168, 174, 176, 177, 185,
    196

Drumlins, 8, 14, 21, 22, 38, 44, 50, 51, 55, 56, 71, 73, 76-7, 97, 120, 122, 144, 147, 148, 149, 155, 168-73, 176, 196, 197, 199, 200, 201, 213, 214, 216, 219, 220, 221

Dublin, 4, 20, 21, 24, 25, 73, 78, 88, 95, 165

Dublin Bay, 4, 11, 24, 78, 98, 113, 183, 185

Dublin Mountains, 11, 15, 25-6, 31, 88

Dundalk, 73, 160, 165, 183

Dundalk Bay, 6, 11, 78, 168, 183

Dundrum Bay, 213, 214

Dungarvan Harbour, 4, 103, 140, 189

Dungarvan Syncline, 105, 107, 118

Dunmore Cave, 41

Dury, G.H., 86, 96, 111

Dwerryhouse, A.R., 118, 152, 154, 167

Early Sperrin Glaciation, 152, 154, 160

East Wicklow Escarpment, 29

Eastern General Glaciation, 115

Enniskerry, 27, 109, 130, 133

Enniskerry Mountain Glaciation, 126, 129

Erne, Lower Lough, 22, 149, 176

Erne, Upper Lough, 22, 149, 176

Erriff valley, 50, 110

Errigal, 14, 60, 64, 65, 84

Erris, 51, 145, 147

Erris till, 147

Eskers, 8, 21, 120, 157, 176-9, 199

Fair Head, 70, 71, 206, 207

Fanad Head, 62, 155, 156, 203

Fanad Peninsula, 60, 62

Farrington, A., 28, 93, 95, 108, 115, 126, 129, 130, 133, 195

Fedamore Moraine, 120

Fethard, 124, 138, 187

Finn river, 111, 154

Flatrès, P., 95

Forth Mountain, 31, 138

Foxford, 55, 110

Foyle basin, 5, 64, 65, 72, 86, 98, 111, 112, 116, 147, 152, 154, 155, 157, 158, 168, 174, 201, 204, 205, 219

Galtrim Moraine, 120, 177, 179

Galty Mountains, 22, 39-40, 136, 143, 144, 145

Galway, 4, 96, 195

Galway Bay, 11, 17, 21, 44, 46, 96, 97, 193, 195, 197

Gap of Dunloe, 38

Garron Point, 72, 208, 210, 211

Garryvoe, 124, 141, 189, 190

Garryvoe till, 141

Geikie, Sir Archibald, 79

George, T.N., 84

Giant's Causeway, 5, 6, 16, 68, 205

Gill, W.D., 86

Gill, Lough, 55, 56

Gilltown Glaciation, 165

Glacial deposition, 8, 21, 31, 51, 54, 55, 56, 66, 71, 77, 91, 97, 112-13, 113-14, 133, 136, 148, 157, 165, 174, 176-9

Glacial drainage channels, 8, 29, 54, 56, 64, 66, 103, 110, 112, 129, 133, 143, 144, 157, 167

Glacial erosion, 8, 28, 38, 39, 44, 45, 48, 51, 52, 54, 55, 59, 63, 64, 66, 71, 74, 91, 103, 110, 111, 113, 114, 148, 155, 201, 213

Glen of Aherlow, 40, 143, 144

Glen of Imail, 28

Glen of the Downs, 29, 129

Glenariff, 71, 72

Glenarm, 72, 210

Glencree, 27, 94, 133

Glendalough, 27, 28, 94, 136

Glendasan, 27, 28, 94, 134, 136

Glengesh Plateau, 86, 123, 155

Glenmacnass, 27, 28, 94

Glenmalur, 28, 94, 108

Glenties, 63, 111

Gort, 124

Gort Lowland, 21, 97

Gortian Interglacial, 115, 123-4, 140, 141, 143, 145, 147, 154, 189

Graiguenamanagh, 94

Graiguenamanagh Gap, 100, 103
Greater Cork-Kerry Glaciation, 141-3
Great Glen Fault, 15, 63
Great Island Anticline, 35, 36, 91, 104
Great Sugar Loaf, 14, 28, 95, 133
Gweestin valley, 7, 82

Hacketstown, 30, 130, 133
Hartley, J.J., 86
Highland Boundary Fault, 15, 55, 65
Hill, A.R., 167, 170-3
Howe's Strand, 124, 189
Howth Peninsula, 125, 183, 185
Hoxnian Interglacial, 115, 123, 189

Iar-Connacht Lowland, 46-7, 95-6, 195
Inishowen Peninsula, 60, 62, 64, 122, 154, 155, 156, 168, 201, 204
Inistioge, 26, 94, 98
Inistioge Gap, 100, 103
Ipswichian Interglacial, 115, 123, 126, 128, 189, 190
Island Magee, 208, 210
Iveragh Mountains, 143
Iveragh Peninsula, 7, 37, 84

Jessen, K., 123
Joyce's river, 50, 148
Jukes, J.B., 88, 101, 102, 104

Kames, 21, 120, 136, 155, 199
Karstic phenomena, 21-2, 35, 41, 42, 44-6, 55, 59, 77, 97, 102, 105, 106
Kells Moraine, 120
Kenmare, 32, 34, 105, 143
Kenmare river, 190
Kerry Head Peninsula, 42, 192
Kidson, C., 190
Kilbeg, 140
Kilkeel, 74, 162, 214
Kilkeel Glaciation, 165
Killadoon till, 145, 147
Killala Bay, 6, 55, 197, 199
Killaloe Gorge, 101, 102, 103, 113-14
Killarney, 5, 38, 42, 81, 143
Killary Bay Little, 50

Killary Harbour, 12, 48, 49, 50, 146, 148, 195, 196
Killary Mountains, 48-51, 52, 87, 90, 95-6, 110-11, 122, 145, 195
Killough, 168, 174, 213
Killough Moraine, 120
Killybegs, 16, 63, 96, 155
Kilmore Quay, 118, 124, 140, 187
Kilrea, 174, 176
Kilrea Moraine, 120
Kilroe, J.R., 152
Kings river, 28, 108, 109, 134
Kingscourt, 5, 7, 86
Kinsale, 37, 91
Kinsale Harbour, 104, 189
Knockadoon Head, 98, 189
Knocklayd, 68, 112
Knockmealdown Mountains, 12, 35, 36, 92, 105

Lagan valley, 5, 72, 112, 167, 174, 176, 211
Laragh, 95, 109, 136
Larne, 68, 82
Larne lough, 208, 210
Late Sperrin Glaciation, 154, 155
Lee river, 35, 98, 103, 104-7, 143
Leinster Axis, 24-31
Leinster Mountains, 24-31, 78, 93-5, 100
Lesser Cork-Kerry Glaciation, 120, 143, 192
Lewis, C.A., 143
Liffey river, 27, 28, 95, 97, 98, 108, 109-10, 113
Limerick, 20, 22, 24, 114
Limerick volcanic basin, 20
Linton, D.L., 12, 13, 81
Lisburn-Dunmurry Moraine, 120
Liscannor Bay, 44, 192, 193
Listowel-Killorglin Lowland, 39, 42
Londonderry, 157, 204, 205
Longford-Down Axis, 1, 13, 15, 17, 72, 213
Lough Neagh Clays, 6, 69, 84, 85, 90, 111
Louisburgh, 50, 51, 96, 147
Louisburgh Cool Phase, 147
Low Plio-Pleistocene sea-levels, 98-9

Lower Bann valley, 69, 70, 112, 157, 167, 168, 174, 176, 205
Lugnaquillia, 15, 24, 25, 26, 29, 30, 31, 93, 108, 136

M1 Motorway, 170
McCabe, A.M., 160, 165
Macgillycuddy's Reeks, 38, 86, 143
McKerrow, W.S., 50, 87, 95, 110
Maghera, 155, 214
Magho Mountain, 84, 148
Magilligan foreland, 204, 205
Main river, 70, 112, 165
Mallin Beg, 116, 147, 152, 201
Malin Head, 156, 201, 203, 204
Mallow, 34, 37, 141
Manorhamilton, 54, 55
'Manure Gravels', 7
Mask, Lough, 17, 21, 22, 48, 87, 148
Maumtrasna Massif, 49, 50-1, 87, 95, 110
Maumturk Mountains, 49, 50, 146
Mell Glacio-Marine Formation, 160
Melvin, Lough, 58
Mesolithic artifacts, 210, 221
Midlandian Glaciations, 8, 69, 115, 116, 118-23, 128, 130, 133, 134, 136, 138, 143, 147, 148, 149, 154, 160, 165, 167, 168, 170, 176, 181, 190, 192, 201
Midleton, 8, 98
Mid-Ulster Highlands, 64-6
Military Road, 26, 94
Miller, A.A., 92, 93, 95, 97
Mitchell, G.F., 116, 123, 130, 133, 138, 143, 145, 179, 210
Mitchelstown, 21
Mizen Head Peninsula, 37
Monavullagh Mountains, 35, 36, 92
Moneydorragh Glaciation, 162
Morris, P., 7
Mothel till, 140
Mourne Mountains, 6, 15, 73-7, 83, 96-7, 112, 118, 122, 136, 152, 160, 162, 165, 214
Moville, 155, 204
Moville Glaciation, 120, 154
Moy river, 110, 199
Muckish, 60, 64, 65, 111

Muckros Head, 155, 201
Mullaghareirk Mountains, 42, 145
Mullet Peninsula, 54, 196, 197
Mulroy Bay, 60, 203
Munsterian Glaciation, 8, 115, 116-18, 128, 129, 130, 138, 140, 141-3, 145, 152-7, 158, 190
Murchison, Sir Roderick I., 9
Murlough Nature Reserve, 214, 221
Murphy, T., 8, 98
Murrisk, 49, 51, 145, 148
Mweelrea Mountains, 49, 50, 51, 146, 147

Nagles Mountains, 92, 105
Nahanagan, Lough, 134-6
Nahanagan Stage, 133, 145
Neagh, Lough, 69-70, 71, 77, 85, 111-13, 116, 122, 149, 152, 154, 158, 160, 162, 165, 167, 168, 170, 176
Neolithic period, 219, 221
Nephin Beg Mountains, 52, 54, 136, 147, 148
Newtown, 124, 125, 140, 189, 190
Newtown till, 140, 141
Newtownards, 73, 171
Nore river, 26, 30, 41, 81, 94, 98, 100, 103, 107, 108
North Cork Syncline, 32, 103
North Star Dyke, 71
Northwestern Mayo, 51-4, 96, 110, 122, 145
Nunataks, 122, 123, 133, 138, 143, 146, 148, 155, 160, 165

Old Head of Kinsale, 32, 189
Omagh, 65, 96, 157
Orme, A.R., 93, 98
Oswald, D.H., 58
Ox Mountains, 12, 15, 54-5, 110, 148

Partry Mountains, 49, 50, 148
Peat, 9, 21, 48, 51, 54, 59, 70, 96, 167
Periglacial phenomena, 39, 48, 54, 71, 118, 120, 124, 125, 126, 130, 138, 140, 141, 145, 149, 154, 155, 157, 171, 176, 179, 185, 190, 193, 197, 201, 203, 205, 214

Pingos, 124, 130, 138
Planation surfaces, 35-6, 37, 48,
    90-8, 195
Poisoned Glen, 63, 84
Portrush, 68, 70
Powerscourt waterfall, 109, 133
Poyntz Pass, 69, 83
'Pre-Glacial' beach, 124-5, 126,
    140-1, 143, 187, 189
Prior, D.B., 122, 167, 211

Raised beaches, 64, 72, 143, 156,
    157, 162, 174-6, 185, 187, 189,
    190, 192, 195, 197, 199, 200, 203,
    204, 205, 210, 214, 216
Rathcormack Syncline, 32, 35, 105
Ree, Lough, 20, 22, 113, 122
Reef-knolls, 20, 56
Ria coastline, 32, 37-8, 99, 192
Ridge and Valley Province, 32-9, 86,
    91-3, 99, 100, 103-7
Roddans Port, 174, 179, 216, 219
Rohleder, H.P.T., 76
Rosses, The, 62, 64, 111, 201
Roundwood Basin, 28, 94, 109

Scalp, The, 29, 129, 133
Scarriff, 113-14
Screen Hills, 31, 136, 138, 187
Shanagarry, 141, 189, 190
Shannon Pot, 59
Shannon river, 1, 17, 21, 22, 39, 42,
    44, 59, 81, 100, 101, 102, 103, 104,
    113-14, 120, 192
Sheeffry Hills, 49, 50, 146
Sheep Haven, 60, 201
Sheep's Head Peninsula, 37
Shehy Mountains, 34, 105
Shelton Abbey, 29, 130
Shortalstown, 115, 123, 124, 126-8,
    138
Silent Valley, 74, 162, 165
Silvermine Mountains, 39-40
Singh, G., 179
Slaney, river, 29, 94, 98, 100, 103,
    107, 108, 109, 136, 187
Slemish Mountain, 70, 96
Slieve Anierin Mountains, 56
Slieve Aughty Mountains, 39-40, 113

Slieve Bernagh Mountains, 24,
    39-40, 101, 113
Slieve Bloom Mountains, 39-40
Slieve Croob, 73, 122, 176
Slieve Donard, 74, 76, 83
Slieve Elva, 45, 46
Slieve Felim Mountains, 122
Slieve Gallion, 65, 66, 176
Slieve Gullion, 6, 73, 77, 78, 83, 84,
    112
Slieve League, 60, 79, 155, 201
Slieve League Peninsula, 60, 63, 64,
    96, 116, 122
Slieve Mish Mountains, 39, 86
Slieve Snaght, 60, 65
Slieveardagh Plateau, 40-1, 42
Slievenamon, 36, 37
Slievenamuck ridge, 40
Slievetooey, 60, 155, 201
Sligo, 55, 56
Sligo Bay, 22, 55, 101, 149, 168, 199
Sligo Syncline, 55
Smith, A.G., 179
Smith, C., 8
Somerville and Ross, 91
South Ireland End Moraine, 113,
    120, 123, 192
South Ireland Peneplane, 91-3, 95
Southern Mountain Inliers, 39-40,
    101
Southern Uplands Fault, 15
Spa, 124, 143
Spelga, 74, 162, 213
Sperrin Mountains, 15, 64, 65-6, 86,
    96, 122, 152, 154, 157, 166, 176
Spincha till, 154
Stephens, N., 210
Strangford Lough, 6, 72, 97, 173,
    213, 219
Submerged peat, 187, 197, 204, 213,
    219
Suir valley, 7, 100, 104
Sutton, 125, 129, 185
Sweeting, M.M., 93
Swilly, Lough, 60, 63, 152, 154, 155,
    157, 168, 203
Synge, F.M., 51, 120, 122, 125, 126,
    129, 130, 133, 134, 140, 143, 145,
    146, 147, 190

Tertiary denudation, 79-84
Tertiary deposits, 5, 6, 7, 8
Tertiary diastrophism, 6, 7, 29, 65, 69, 70, 77, 81, 84-90, 102, 107, 108, 111, 201
Tertiary Igneous Mountains, 73-8, 158, 160, 165
Tertiary intrusions, 6, 7, 48, 52, 63, 73-8, 84, 86, 201
Thorp, M.B., 96
Tinahely Hills, 29, 30, 100, 108, 109, 110
Tipperary Moraine, 120, 136, 165
Tors, 27, 29, 76, 138, 155, 201
Tralee, 42
Tralee Bay, 145, 192
Tramore, 26, 31, 92
Tramore Bay, 91
Trim, 120, 176, 177, 179
Tullamore, 176-8
Tullow Lowland, 15, 26, 29, 30, 31, 94, 100, 102, 108, 109, 110
Turloughs, 21, 46
Twelve Pins of Connemara, 49, 50

Vale of Arklow, 29, 94
Vale of Avoca, 28, 94
Vale of Clara, 94, 95

Vernon, P., 173

Walsh, P.T., 81, 82, 86
Warren, W.P., 190
Waterford Harbour, 101, 187, 189, 190
Watergrasshill Anticline, 36, 145
Watts, W.A., 81, 123, 179
Wexford, 31, 98, 100, 126, 136, 138, 187
Whiting Bay, 125, 190
Whittow, J.B., 81, 107
Wicklow, 128, 129
Wicklow Gap, 109
Wicklow Glens, 27-8, 94-5, 108
Wicklow Head, 26, 29, 125, 126, 129, 130, 183, 185, 187
Wicklow Mountains, 9, 15, 24-31, 90, 107-10, 113, 118, 120, 122, 126-38, 145
Williams, P.W., 12, 45
Wilson, R.L., 84
Wolstonian Glaciation, 8
Wright, W.B., 99

Yeats, W.B., 59
Youghal, 8, 98, 103, 141